W9-BFZ-764

THE OCEANS

Other Books by Ellen Prager:

Furious Earth: The Science and Nature of Earthquakes, Volcanoes, and Tsunamis

Other Books by Sylvia Earle:

Sea Change: A Message of the Oceans

Wild Ocean: America's Parks Under the Sea

THE OCEANS

Ellen J. Prager

with Sylvia A. Earle

McGraw-Hill

New York • San Francisco • Washington, D.C. • Auckland • Bogotá
Caracas • Lisbon • London • Madrid • Mexico City • Milan
Montreal • New Delhi • San Juan • Singapore
Sydney • Tokyo • Toronto

Library of Congress Cataloging-in-Publication Data

Prager, Ellen J.
The oceans / Ellen J. Prager with Sylvia A. Earle.
p. cm.
Includes index.
ISBN 0-07-135253-8 (acid-free paper)
1. Ocean. 2. Oceanography. 3. Marine biology. I. Earle, Sylvia A. II. Title.

GC11.2 .P74 2000
551.46—dc21 00-021141

McGraw-Hill
A Division of The ***McGraw-Hill*** *Companies*

A portion of the material in "The Geological Ocean" previously appeared in *Furious Earth,* McGraw-Hill, 1999.

1 2 3 4 5 6 7 8 9 0 DOC/DOC 0 9 8 7 6 5 4 3 2 1 0

ISBN 0-07-135253-8

This book was set in Perpetua by North Market Street Graphics. Printed and bound by R.R. Donnelley & Sons Company.

This book is printed on recycled, acid-free paper containing a minimum of 50% recycled, de-inked fiber.

To a wonderful cadre of mentors, colleagues, and good friends who have taught me much, shown me more, and endlessly provided encouragement and inspiration. This book is a testament to your generosity of spirit and ability to share and foster a passion for ocean science in others. Among the many are several Bobs, Gene, Betsy, Bill, John, Denny, Hal, the folks at SEA, Harry, Greg, Peter, and, most especially, Linda and Sylvia.

Contents

Foreword

The Oceans: Cornerstones of Life

The ocean has sustained and shaped our existence on Earth since life first emerged from the ancient sea. Today, nearly three-quarters of the planet are covered by ocean, 97 percent of the Earth's water lies in the sea, and within it lies 97 percent of the planet's living space. Air-sea interactions dictate climate and weather, and ocean currents regulate the Earth's global thermometer. Spreading of the underlying seafloor creates towering mountains, deep undersea trenches, and active hydrothermal vents that abound with strange marine organisms. The ocean teems with life millions of years in the making, from the tiniest of microscopic bacteria to the largest of living creatures, the blue whale. And, from the ocean's bounty of small floating plants, our atmosphere is thankfully enriched with oxygen and depleted of carbon dioxide. In short, without the ocean there would be no Earth or life as we know it, for ours is a life-giving, water-blessed planet.

But the future of the ocean is far from clear; once considered an impervious system, it now exhibits global signs of degradation. Species once thought to be infinitely resilient, no matter how many were removed from the sea, are proving to be just as vulnerable to decline as buffalo, passenger pigeons, elephants, and tigers. Fisheries including swordfish, cod, tuna, and salmon have collapsed or on the verge of doing so, and many of our coastal regions are polluted, overdeveloped, or eroding. Whales and dolphins are increasingly washing ashore dead or dying. Coral reefs, one of the most spectacular and diverse environments on the planet, are showing global signs of ill health, yet some continue to be fished ruthlessly with cyanide, bleach, and dynamite. And, a marine "dead zone" has developed off the Mississippi delta in the Gulf of Mexico, and may mean the permanent destruction of an underwater area the size of New Jersey. The key to preventing further degradation of the ocean and protecting—maybe even restoring—it for future generations lies in our understanding of the sea and our ability to manage our impact on it.

Over the last century, advances in marine technology and research have enabled us to make great progress in understanding the ocean and its influence on the planet. In 1997 and 1998 an intense climate event known as El Niño wrought havoc on life and property throughout the world and dramatically illustrated the importance of ocean-atmosphere interactions. Using a series of monitoring buoys in the equatorial Pacific, satellite imagery, and computer models, scientists were able to accurately predict the onset of El Niño and begin to forecast its effects. There is evidence that El Niño has been occurring for hundreds, thousands, and possibly millions of years; but 1997 was the first time ever that we were able to see it coming. The anticipated and actual impact of El Niño received widespread media coverage; and El Niño became a common household word—the poster child for all weather-related woes—yet in many areas of the world people were still unprepared for the intensity of El Niño–related drought and flooding. Today, scientists continue to study El Niño to determine why it occurs when it does and why some years are more intense than others. Our growing understanding of how the ocean interacts with the atmosphere will help not only to predict such events but also to better forecast and model hurricanes, global climate change, and sea-level rise.

In the 1950s the largest mountain range on Earth, the midocean ridge system, was discovered beneath the sea, and soon the theory of seafloor spreading was confirmed. These monumental scientific achievements were pivotal to widespread acceptance of plate tectonics, a concept that has revolutionized our understanding of the planet and the evolution of its surface and its life-forms. Over a decade later, scientists discovered an entirely new and thriving ecosystem where superheated plumes of mineral-rich water emanate from fractures in the deep seafloor. Massive colonies of tube worms red with hemoglobin, giant clams, blind shrimp, and an assortment of other bizarre and previously unknown marine creatures were discovered. Most surprising of all was that unlike all other ecosystems on Earth, which are founded on energy derived from the sun, life at the sunless deep-sea hydrothermal vents flourishes on bacteria that use only heat and chemical compounds to grow. In addition, researchers have now found that the circulation of seawater through deep-sea vents plays a surprisingly significant role in the ocean's chemistry and geology. Exploration of the ocean floor using sophisticated acoustic, drilling, and other sampling techniques is revealing ever more about the watery world we live in and its evolutionary past.

Today, as never before, humans can also enter the sea and observe its creatures on their own terms. No longer must we simply drag a net behind a ship to investigate the ocean's biology—although this is still done. Now we can use SCUBA gear (self-contained underwater breathing apparatus), small manned submarines (submersibles), or unmanned vehicles operated from above (remotely operated vehicles). Scientists are just beginning to shed light on the amazing world of ocean life. We are learning about the sea's twilight region where creatures commonly create dazzling displays of bioluminescent light, make a daily commute to the surface at dusk, and are often little more than a wealth of gelatinous goo that nature has fashioned into exquisitely delicate forms. At the sea's surface we are gaining insights into the fascinating world of marine mammals, sharks, and the majestic tuna—elite among the sea's swimmers with a physiology to make athletes drool. On the seabed, scientists are learning that seemingly benign creatures, like starfish, sea cucumbers, and some shrimp and shellfish, are cunning predators, equipped with highly specialized grazing and hunting tools and stunning means of defense. And in the very deepest sea, we are discovering unexpected forms of life and fantastic seascapes where organisms grow in an abundance greater than ever imagined.

Though we have learned so much in the last century, we have in fact explored less than 5 percent of the world's ocean. In the untraveled regions of our deep sea, creatures longer than busses—giant squids—continue to elude inquisitive scientists. Estimates of the ocean's undiscovered species range from as few as 1 million to more than 50 million. We can barely imagine the unknown multitudes that lie hidden in the underwater realm. Why does so much of the ocean remain a mystery even though we have been studying it for more than a century?

Imagine flying in a hot-air balloon over the lush, green canopy of a rainforest. Through the clouds and mist you can just barely make out the treetops and a few of the birds flying among them. What lies hidden in the undergrowth? How many organisms are there, what do they look like, and how do they behave? Using a rope and bucket you blindly drag the rainforest from above hoping to ensnare some of its inhabitants or the materials that make up its infrastructure. But alas, with such feeble and limited means you can learn little about the environment and life below. For years this is essentially how we have studied the ocean—blindly sampling the sea with limited and relatively ineffective methods. Even today, with technology as advanced as it is, study of the ocean remains a difficult and expensive task. Whether through large-scale satellite imagery, small-scale chemical and biological measures, or even the collecting of fossil impressions of ancient sea creatures, all aspects of oceanographic study require some type of observation or sample collection, and herein lies the problem.

Simply, the ocean is big, wet, cold, dark, and inhospitable to air-breathing terrestrial creatures like ourselves. On land, scientists can observe and collect samples while walking, driving, or flying from one environment to the next, sometimes crossing several ecosystems in less than a few hours. For example, in the coastal zone it is often possible to walk from a forest across a marsh to a beach, easily traversing three distinct ecosystems. In the ocean, ecosystems tend to be large, hundreds or thousands of kilometers in scale. It can take days to weeks for a ship to move from one environment into the next. Although satellite technology now allows us to capture large-scale images of the ocean, they reveal only the very surface at one moment in time—essentially snapshots of the sea's thin skin. And, the ocean is truly a three-dimensional world, because environments range not only across the sea, but also down with depth.

For humans to actually enter the sea and spend any length of time observing and sampling its deeper realms, we must bring warmth, air, artificial light, and some means of propulsion to move through the water. We must also protect ourselves from the pressure, the great weight of the overlying water. No wonder exploration of the sea remains one of the most technologically difficult and expensive tasks on the planet. But humans are ingenious creatures, and over time we have developed clever ways of sampling and visiting the deep-ocean realm. Unfortunately though, we are still relegated to brief visits, restricted in both time and space.

Consider this idea of sampling a bit more carefully. For instance, if you could take just 10 samples of the seafloor to determine its nature, where, when, and how would you do it and what would the results mean? If 1 out of 10 of the samples was hard black rock, does that mean that 10 percent of the seafloor is made up of such rock? And if so, how is it distributed over the wide seafloor? For any scientific endeavor the number of samples collected greatly influences the magnitude of the knowledge gained. The more samples collected, the more accurate the interpretation of the data. Because data collection in the ocean is a time-consuming and expensive endeavor, often we can take only a few samples. Consequently, much of our understanding is educated guesswork, extrapolated from limited data, and often the first interpretation is later proven incorrect as more observations are made. Not only can the number of samples influence our interpretation; so can our collection methods. Scientists collecting samples from the sea must design devices that retrieve unbiased specimens or are specifically equipped for a particular purpose. The diameter of holes in a net must be considered when collecting fish or plankton, the size and density of floats must be carefully calculated when tracking currents, and the hardness of the seafloor must be contemplated when attempting to obtain sediment samples. One would not try to grab a small copepod or fish with a submersible's awkward and large mechanical arm or use a suction tube to capture a mighty tuna.

While many people are concerned about what humankind is putting into and taking out of the sea, the greatest threat to the future of the ocean—and thus to ourselves—may be our own ignorance, what we still don't know. Today, ocean scientists work together with technicians, economists, computer specialists, and experts from many other fields to better understand the sea and how it influences everyone, everywhere, everyday. We are beginning to recognize the complexity of ocean-climate interactions, the nature of the underlying seafloor, the true diversity and abundance of organisms in the sea, and the links between species, habitats, and environmental disturbance. As we become more knowledgeable about the ocean and appreciate the connection between what happens in it and our own lives, no matter where on Earth we live, we will be inspired to take care of the sea, which takes care of us.

In this volume, my colleague, friend, and sometime diving partner, scientist Ellen Prager, provides an overview of our current understanding of the oceans and a compelling case for the need of a greater "sea" of knowledge. Few are as well qualified as Dr. Prager to take the reader on a watery journey through the ocean's vast and wide-ranging wonders. No one is better at conveying concepts and translating the arcane language of scientists in a way that make sense to everyone, with an added spice of wonder—and humor—that is simply irresistible.

Sylvia A. Earle

Preface

THE OCEANS ARE a fascinating and vital part of our world. They have inspired myth and legend, classic works of art and literature, and numerous television shows and movies. Today, we also recognize the importance of the sea as an integral and sustaining part of our planet, yet most of us know little about what science has discovered, or has yet to discover, about the ocean's evolution, its flow, interactions with climate, underlying rock and sediment, or its many wondrous and strange creatures. For decades, dedicated scientists have focused their efforts on the ocean, but the rewards of their research rarely reach the public purview.

As a student, researcher, teacher, explorer, and passionate observer of the ocean, I have been afforded access to many amazing aspects of the sea, often firsthand. Some of my most memorable times, and much of what I know about the ocean, come simply from spending time in and on the water and talking and working with an array of marine scientists, technicians, ship crews and students. The greater public is rarely privy to much of what I have been lucky enough to see, hear, and experience, or such information is often explained in a way that is difficult to understand. Yet nearly every day I meet someone outside the realm of science who is fascinated by the ocean, once dreamed of being a marine scientist, and wants to know more about the sea. This book is meant for them. It is not meant to be a comprehensive scientific discourse on the ocean; rather, it is a means of sharing many of the wonders scientists have learned over the years about the sea and its creatures.

Someone once told me that writing a book was the hardest thing he had ever done. Yet writing *The Oceans* has been a labor of love and a thing of joy. In the midst of writing, I would often excitedly call friends, colleagues, and relatives and implore them to listen to a passage I had just typed. Most often, they too were enthusiastic in their response, though it may have been an attempt to simply humor me. And while it is somewhat of a relief to have completed the book, in many ways I am sad to see it done. I hope that those who read this book, young or old, will share my passion for the ocean, understand it a bit better, and be inspired to delve even farther into its delightful depths.

Dr. Sylvia Earle, world-renowned ocean explorer and conservationist contributes to the foreword, afterword, and with inspired words throughout the book. With great vision towards the future and an unending love for the sea, Dr. Earle is one of the ocean's most accomplished, outspoken, and admired advocates.

Great thanks go to those who provided assistance with research materials and photographs and to all who encouraged me by their support and inspiration. Particular thanks to Linda Glover for review and comments on ocean policy, Admiral James Watkins for our discussions on the future of ocean science, Gene Shinn and Billy Causey for their fond recollections of the Florida Keys in times past, and Bob Halley for a wealth of information and advice. Appreciation is extended to Pamela Baker, Steve Gittings, Wolcott Henry, Larry Madin, Steven Miller, Frank Muller-

Karger, Vita Pariente, Raymond Rye, and Robert Tilling for their assistance and photographs. Thanks to Skye Herzog, my agent, for her good humor, excellent advise, and unwavering encouragement.

My sincere gratitude also goes to the editors and production staff at McGraw-Hill, especially Griffin Hansbury for his enthusiastic support of the project and editorial work. And finally, last but not least, I would like to thank Sylvia Earle for her wonderful contributions to the book and her incredible generosity. She has shared with me her thoughts and dreams, and encouraged me to follow my own.

Ellen J. Prager

In the artificial world of his cities,
[man] often forgets the true nature of his planet
and the long vistas of its history, in which the existence
of the race of men has occupied a mere moment of time.
The sense of all these things comes to him
most clearly in the course of a long ocean voyage,
when he watches day after day the receding rim
of the horizon, ridged and furrowed by waves;
when at night he becomes aware of the earth's rotation
as the stars pass overhead; or when, alone in this world of water and sky,
he feels the loneliness of his earth in space.
And then, as never on land, he knows the truth
that his world is a water world, a planet dominated
by its covering mantle of ocean, in which the continents
are but transient intrusions of land
above the surface of the all-encircling sea.

—*Rachel Carson,*
The Sea Around Us

OCEANS OF THE PAST

Going into the ocean is like diving into the history of life on Earth.
—Sylvia Earle

Evolution's Drama

THE EVOLUTIONARY TALE of Earth and its oceans is undoubtedly the world's greatest adventure, soap opera, and disaster movie all wrapped into one. The setting is the planet Earth, and the characters are all forms of ancient and modern life. The story begins in an exotic and unfamiliar terrain that is often rocked by cataclysmic asteroid impacts, fiery volcanic eruptions, and huge earthquakes that rip apart the land. At times, cold temperatures plunge Earth into a deep freeze, while at other moments the planet is a warm and cozy place to live. Sea level will rise and fall, while the continents shift, crash together, and separate through the ages. The characters of this epic adventure will also change; sometimes they are familiar in form and at other times they are alien creatures of nightmarish proportion. Events occur that cause biologic catastrophes and give rise to new characters, as the old are mortally wounded or usurped by a new ruling class. But throughout much of the story, one factor will remain—the presence and life-giving nature of the sea. The ocean and its inhabitants play a fundamental role in Earth's development and the evolution of life. As humans, we represent a mere instant in time near the end of the story. But by following the evolution of the ocean and life, and the changing nature of Earth, we gain a tremendous understanding of our dynamic planet, the fragility of life, and our own marine origins. This is one story everyone should know, for it truly puts in perspective our own relatively insignificant existence and yet, our potentially enormous impact on Earth.

Before we begin our voyage back in time, it is important to note a few things about how Earth's history has been pieced together and why, like an old puzzle, some of the pieces are missing. Modern technology allows us to record our children's development on video so that one day they may witness their own birth and growth. Unfortunately, there is no recording of the birth and evolution of Earth and its oceans, so scientists must reconstruct their history from clues in ancient rocks, fossils, and other planets. For instance, our understanding of Earth's birth is reconstructed principally from studies of planetary collisions, meteorites, ancient craters, and inert gases such as xenon, krypton, and argon, which are an abundant component of the sun, but rare on Earth.

The difficulties of sampling in the modern ocean seem inconsequential when compared to the problems scientists encounter while studying the ancient Earth, the oceans, and early marine life. Logically, the best information about the early oceans and primitive marine life should come from the sediments and fossils of the seafloor. Unfortunately, as we will see when discussing the ocean's geology, seafloor spreading and the destruction of ocean crust by subduction at deep-sea trenches results in the continual recycling of the ocean floor. Although Earth is billions of years old, the oldest sediments and rocks in the modern ocean are only about 180 million years old. Luckily, continental landmasses are not typically recycled and mountain rocks often contain ancient ocean sediments and fossils that have been lifted high above the sea. But the rock and fossil record is far from complete, preservation is often sparse, and interpretation can be difficult.

For centuries, scientists have searched the continents, the ocean, and even outer space for pieces to Earth's evolutionary puzzle. In this account of planetary evolution, some of the fossils and rocks used to decipher Earth's history are described; however, many more will be neglected for lack of space. The interested reader is referred to the suggested readings at the end of this book for more detailed accounts of the fossil, rock, and sediment evidence used to reconstruct Earth's history.

Another invaluable source of information that will be briefly mentioned, but deserves greater attention is the Deep-Sea Drilling Program, and its successor, the Ocean Drilling Program. Both programs represent an unprecedented international collaboration of scientists, technicians, and managers whose goal is to collect sediment and rock cores from deep beneath the sea. Deep-sea cores have provided some of the most significant and prolific scientific data ever on topics such as plate tectonics, sea-level history, and global climate change.

Earth is some 4.5 billion years old and its evolution has occurred on time scales that are on the order of billions, millions, and hundreds of thousands of years. However, we tend to think of time in terms of a human lifetime, on the order of a hundred years, broken into intervals of years, months, weeks, hours, and minutes. When geologists began studying Earth's history through its rocks, they recognized the need to have a way to reference time in the same terms as Earth's evolution, so they developed the geologic time scale (Figure 1).

Intervals within the geologic time scale are based on the occurrence and disappearance of certain fossils or groups of fossils. Initially, geologic time was divided into time before life and time after life. However, this system was altered because in the past several decades it was discovered that primitive life began much earlier than previously believed. According to the current classification, the earliest period of Earth's history, from 4.5 billion years ago (bya) to about 550 million

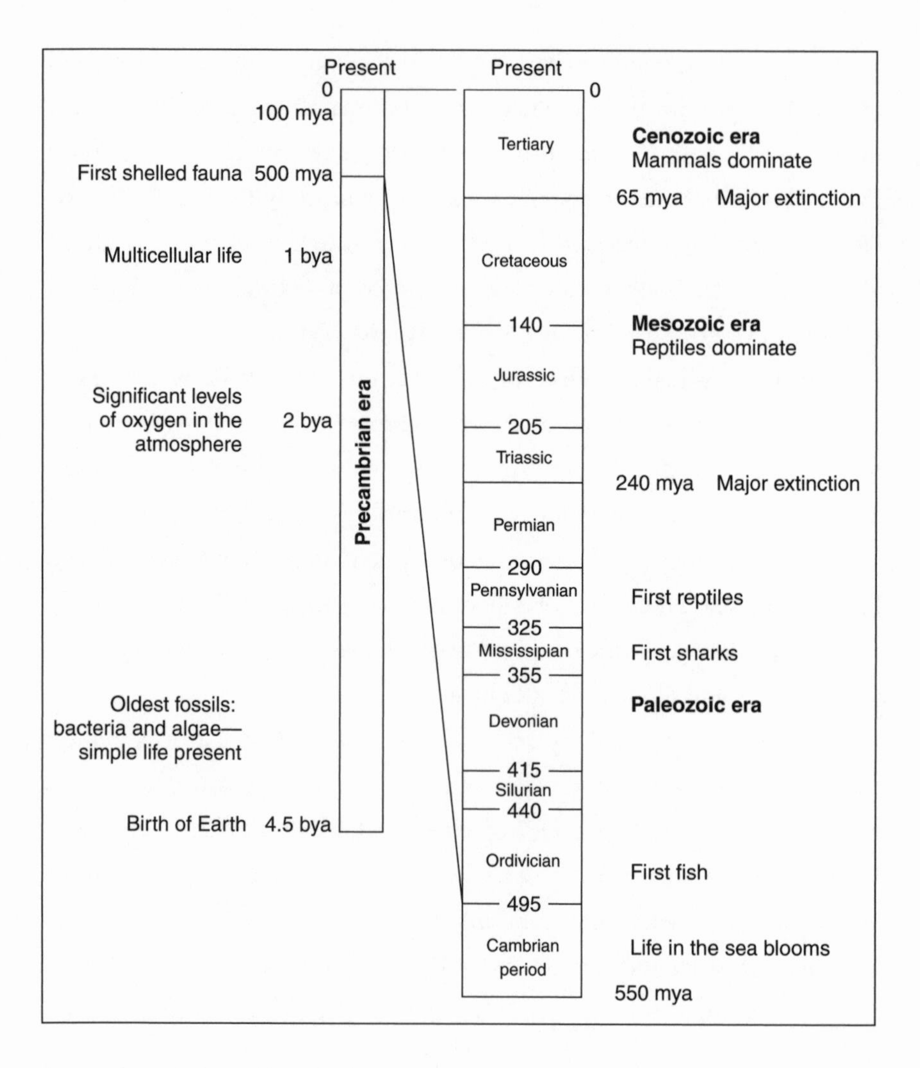

Figure 1 Geologic Time Scale.
(McGraw-Hill.)

years ago (mya), is called the Precambrian*; it represents time before organisms evolved enough to leave abundant fossils. Starting in the Cambrian (≈ 550 mya), periods are grouped into the Paleozoic, Mesozoic, and Cenozoic eras, representing ancient, middle, and recent life, respectively.

The story described here is the latest rendition of Earth's evolution based on decades of exploration, scientific research, and accidental discovery. Undoubtedly, in this grand epic, new details will emerge and new interpretations of the data will be made, but for now, this seems to be the best fit of the evolutionary pieces in the planetary puzzle.

* Many scientists now divide Precambrian time into the Hadean (4.5 to 3.8 billion years ago), Archean or Archeozoic (3.8 to 2.5 billion years ago), and Proterozoic (2.5 to 0.55 billion years ago). The absolute dates that divides the intervals in the geologic time scale are inherently inexact due to the nature of dating methods. Consequently, various sources use slightly different numbers. For instance, the Cambrian Period and Paleozoic era begin at 544, 545, 550, 560, or 570 million years ago depending on the source. Throughout the remainder of the book, the numbers provided as time subdivisions are not meant to reflect an absolute age, but an approximate date based on the sources listed in the reference section.

The Earth and the Ocean Are Born

OUR STORY BEGINS some 4.5 billion years ago in a galaxy called the Milky Way. A cloud of dust and small planetary bodies orbit an early sun. Soon these circling rocks, ranging in size from tiny dust particles to huge asteroids hundreds of kilometers across, begin to collide. At first, the collisions are slow, and gravity causes the crashing space rocks to fuse together. A mass of rock is created—the embryonic Earth. More and more of the colliding rocks aggregate, and Earth grows larger. As it grows, Earth's gravitational attraction becomes more powerful. The orbiting space rocks are pulled in faster and faster, hitting Earth's surface with ever-greater violence, creating huge craters and releasing massive amounts of heat. The heat is so intense that it melts the outer layers of the growing planet, and a shallow sea of boiling molten rock is formed. Large amounts of heat are also trapped inside the growing planet, buried by tons of accreting rock. Over millions of years the bombardment continues and the planet grows until reaching its modern size.

Throughout the period of Earth's early growth, colossal space-borne impacts occur on a regular basis and eject tremendous amounts of dust and ash into the air, blocking out all sunlight and throwing the planet into utter darkness. Comets, huge chunks of frozen gas and ice, and asteroids pelt the planet, and strong winds blow over the surface. Water and gas trapped within the accreting rocks are released into the atmosphere by massive explosions during impacts and through ongoing volcanic eruptions. The atmosphere is harsh and heavy, composed of carbon dioxide, water vapor, nitrogen, and several other gases. Dark clouds of dust, steam, and ash blanket the sky, thunder and giant bolts of lightning pierce the darkness, and a fiery ocean of lava boils and swirls about the surface. The early dark Earth gives no hint of the blue orb that it will become.

How do we know that all this took place some 4.5 billion years ago? Scientists use an ingenious technique to estimate the timing of Earth's birth: *radioactive dating.* All elements on Earth are characterized by a specific atomic weight based on the number of neutrons and protons in their nucleus. Some elements, such as uranium, radium, potassium, and carbon, have several forms, each with a different number of neutrons in their nucleus; these are called isotopes. Isotopes are chemically identical, but have different atomic weights, and some are unstable or radioactive. Radioactive isotopes decay at a specific rate, known as a *half-life.* The half-life of an isotope is the amount of time it takes for the isotope to decay by one-half of its original amount. If geologists know what the half-life of an isotope is, they can calculate how old a rock is by measuring the amount of the parent material and the amount of the daughter product (a product of its decay). For instance, carbon has three isotopes: two that are stable (carbon 12 and carbon 13) and one that is unstable, or radioactive (carbon 14). When carbon 14 decays, it emits energy and produces nitrogen 14. Its half-life is 5570 years; in other words, it takes 5570 years for half the C^{14} in a given substance to decay to N^{14}. Geologists can estimate a rock's age by measuring the amount of C^{14} and the amount of N^{14} that are present. This is known as *carbon dating.*

Radioactive dating of meteorites, which are all believed to have formed at about the same time as Earth, indicates that they are all about 4.5 billion years old. Additionally, the moon is now believed to be a piece of early Earth ripped from its mass when a huge asteroid hit the growing planet, and lunar rocks have all been dated at just younger than 4.5 billion years in age (Hartmann and Miller, 1991). Dating of rocks in the remains of ancient craters, particularly on the moon's surface, suggests that by about 4.5 billion years ago, Earth had grown to its present size and the rate of bombardment by comets and asteroids had begun to slow.

It is now 4.4 billion years ago. Fewer impacts stir and heat the lava ocean, and Earth's surface begins to cool. As it cools, a thin, black crust of hardened lava encases Earth. Although impacts and volcanic eruptions periodically rip apart the crust and spew fountains of fiery lava skyward, collisions are fewer, cooling continues, and a thicker rocky crust forms over the planet's surface. Cooling causes water vapor in the atmosphere to condense into liquid and droplets rain down. Soon torrential rains douse the planet, and the first watery ocean is born. The early ocean is acidic and extremely hot, possibly on the order of 100°C (212°F). Elements provided by volcanic eruptions and outpourings make the primordial sea only slightly salty. A weighty, caustic atmosphere rich in carbon dioxide still surrounds Earth, but as more water condenses sunlight begins to shine through the dark clouds. The tall, jagged edges of impact craters stand out above the sea, yet the forces of erosion are fierce; raging floods carve out deep canyons and wear down the rugged peaks. The last few asteroid impacts generate towering waves, tsunamis that sweep across the planet. And because the moon is closer to Earth, mammoth tides surge through the sea.

Carbon dioxide from the atmosphere begins to dissolve in the new ocean and combine with carbonate ions to form calcium carbonate, or limestone. As more and more limestone is deposited on the seafloor, carbon dioxide is increasingly removed from the atmosphere and the skies brighten. Like an antacid acting on an upset stomach, the calcium carbonate buffers the sea's acidity, making it chemically less harsh. Incoming radiation from the sun increases and the warm temperatures cause massive amounts of water to evaporate from the early ocean. Sea level falls and much of the underlying surface is exposed. Weathering of the newly exposed land by rain and rivers brings more minerals into the ocean and its saltiness increases.

During this period climate variations may have been extreme, while volcanic eruptions, earthquakes, and tsunamis continued to ravage the planet's surface. Some scientists believe that catastrophic asteroid impacts continued throughout this time and the global ocean was periodically vaporized, only to reform in the following decades.

Life Begins

IT IS NOW 3.8 billion years ago. The era of intense bombardment from space has ended and the restless Earth emerges as a blue planet, its surface covered by an azure sea freckled with barren rock islands. On land and beneath the sea, dense, dark basaltic rock, a fine-grained silicate rock rich in iron and magnesium, forms a thick crust, scattered with "floating" blocks of lighter granitic material—light-colored, silicate rocks rich in potassium, calcium, sodium, and aluminum. (These granite "rockbergs" will eventually thicken and form the nuclei of our continents.) The sky is brightening, the atmosphere is thinning, and the climate is slowly cooling, but still no plants or animals inhabit the land or sea.

When and how life began on Earth is one of the most intriguing and hotly debated questions of all time. Is the ocean, 4 billion years ago, a primordial soup, thick with organic molecules? And if so, where does the first organic matter come from? Some suggest that organic materials, the building blocks of life, are brought to Earth from space in asteroids or comets. Others believe that these substances are generated within

Earth's ancient sea itself. But regardless of where the organic building blocks come from, life begins in the ocean.

Clues to what Earth looked like at this time and the nature of the first organisms are found in ancient sediments that are now hardened rocks on land. So far, the oldest sedimentary rocks found on Earth are some 3.75 billion years old and come from the Isua Hills of Greenland. Discovered in 1971, the Isua Hills deposit contains a sequence of fine-grained rocks and dark, hardened lava in strange tubular and pillow-shaped forms, like crusty black toothpaste oozed from a tube. These odd-shaped rocks are called pillow basalts; they form when molten lava erupts on the seafloor and is instantly cooled by cold seawater (Figure 2). Ancient pillow basalts have also been uncovered in the rocks of the Barberton Greenstone Belt in southern Africa. Other rocks have been found with surfaces that look amazingly like a bubbling mud pool that has solidified. Today, hot baths of simmering mud are common in geothermally active regions, such as Yellowstone National Park. Similar rocks dating 3.2 to 4 billion years old have been found in Australia and northern Canada. But the most amazing discovery of all was in South Africa. In rocks of hardened silica, called chert, geologists discovered distinct microscopic, rice grain–shaped fossils. It is believed that these are the preserved remains of primitive bacteria that once lived in ancient steaming muds. The idea that heat-loving microscopic life began in bubbling pools of mud or possibly in volcanically active undersea areas seems plausible given recent revelations in the deep sea.

In 1977, geologists discovered new and unusual forms of marine life living at deep-sea hydrothermal vents along the Juan de Fuca ridge off the Seattle coast. At more than 2500 meters below the sea surface, giant clams, tube-dwelling worms (tube worms), crabs, and other peculiar marine creatures were found clustered around plumes of hot water emanating from fractures in the seafloor. One of the most astounding aspects

Figure 2 Underwater photograph of pillow basalts on the Hawaiian submarine volcano Loihi.
(Courtesy of the U.S. Geological Survey.)

of the deep-sea vent discoveries was that the marine organisms at this and other sites discovered later, were feeding on or living off of chemosynthetic bacteria. *Chemosynthesis* is the process by which organisms use heat, water, and chemicals, such as hydrogen sulfide, to create organic matter. In contrast, photosynthesis is the process by which plants use the sun's energy, water, and carbon dioxide to produce organic material and oxygen. In the vast majority of Earth's ecosystems, photosynthesis is the underlying engine that drives life. The discovery that chemosynthesis was the base of a flourishing food chain deep beneath the sea was astonishing to scientists throughout the world, and opened the possibility that life did not begin at the ocean's surface, but within its deepest reaches, at sites of deep-sea hydrothermal activity. Today, we know that chemosynthetic bacteria can proliferate in the deep ocean and within other environments once considered hostile to life, including the famous hot springs and mud pools of Yellowstone National Park, and in natural oil or gas seeps in the deep Gulf of Mexico (Color Plate 1). But it is still not clear where life begins. Do small bacteria thrive on the heat generated by Earth in hot springs, bubbling mud pools, or deep-sea vents, and then later move into the shallow sea to harness the powerful energy of the sun?

It is now about 3.2 billion years ago, and Earth remains a very hostile environment. Hot lava issues forth on land and beneath the sea, boiling hot springs typify the landscape, and the atmosphere is still relatively thick with steam and carbon dioxide. However, simple single-celled life is beginning to take hold.

In the Fig Tree formation rocks of Australia, dated at 3.2 billion years old, geologists have found large rod and spherical-shaped fossils. These fossils resemble modern photosynthesizing bacteria and blue-green algae, now called *cyanobacteria*. Fossil algae and bacteria-like forms were also found in the rocks of the Gunflint Chert, a 2 billion-year-old deposit along the shores of Lake Superior in western Ontario. But here, geologists noted that fossils formed odd dome- and pillar-like layered structures. These peculiar forms appeared to be biologically derived, but their origin remained a mystery for many years. Then, similar stubby, pillar-like colonies of cyanobacteria were discovered growing in the shallow tide pools of Shark Bay, Australia, and larger structures were recently found in the shallow tidal channels of the Bahamas (Figure 3). These impressive, living columns, called *stromatolites,* can grow up to several meters in height or width. Stromatolites form as cyanobacteria or algae grow upward and produce sticky, fibrous layers of organic material that are periodically overlain with sediment. Sometimes a cement-like coating of calcium carbonate is also produced. Once grazers evolve, stromatolites will exist only in environments where currents, high salinity, periodic drying, or other conditions exclude the munching of under-

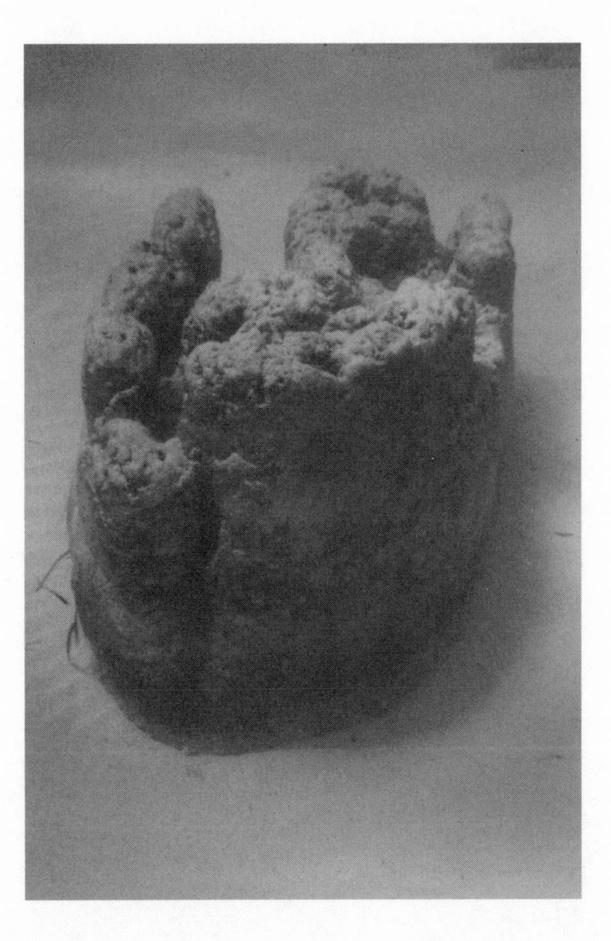

Figure 3 Underwater picture of a stromatolite growing in a tidal channel off Lee Stocking Island, Bahamas. (E. Prager.)

water creatures. However, prior to the evolution of such creatures, stromatolites are abundant. Some have been dated at older than 3 billion years, confirming that life in the shallow sea is dawning.

It is some 3 billion years ago. As the skies clear and Earth slowly cools, the planet's surface subtly begins to change. Volcanoes continue to spew fire and ash, but now extensive shallow-water areas and boiling pits of mud are infested with bacteria and early algae. Tide pools are covered by a blue-green slime of life and fields of stromatolites dot the horizon (Figure 4). Bacteria also flourish in the deep sea at sites of hydrothermal activity. Limestone deposition and newly photosynthesizing organisms continue to lower levels of carbon dioxide in the atmosphere, and Earth's climate cools further.

Carbon dioxide in the atmosphere absorbs radiation (heat) emitted from Earth's surface. Increasing levels of carbon dioxide increase heat absorption and cause the climate to warm: the so-called *greenhouse effect.* Scientists think that this process operated in the same way on the young planet. However, decreasing levels of carbon dioxide caused Earth's climate to cool rather than to warm (Earlier decreases in carbon dioxide are thought to have been offset by increased solar radiation.)

Figure 4 Earth as it may have been some 3 billion years ago.
(Courtesy of the Smithsonian Institution.)

Earth's first life-forms are microscopic, single-celled organisms. But then one of the most controversial and mysterious stages of evolution occurs: multicellular life evolves. Organisms obtain cells that have a nucleus and specialized intracellular structures. Did they simply evolve from existing single-celled organisms? Or, given the symbiotic nature of the structures within cells, did small, single-celled organisms and molecular bodies coalesce into multicellular life-forms? Either way, multicellular marine life evolves sometime between 2 and 3 billion years ago. No one is really certain when and how it happened, but fossils and rocks indicate that a key factor in the evolution of multicellular life is the appearance of free oxygen in the atmosphere.

Until 2 to 3 billion years ago, carbon dioxide and water dominated Earth's atmosphere because there was no means of producing significant amounts of oxygen. But at some point, oxygen from early photosynthesizing organisms began to accumulate in the atmosphere; more was produced than used. The rustiness of ancient sediments provides a clue to the evolution of oxygen in the atmosphere. Oxygen is a highly reactive gas, and when it combines with iron it creates rust. Before oxygen was a significant component of the atmosphere, dark iron-rich sediments eroded from the land and were transported into the ocean. Over time, these sediments collected on the seafloor, were buried, and eventually hardened into rocks. Throughout the world, rocks dated at 3.8 to 2.3 billion years in age are composed of alternating dark, iron-rich layers and light, iron-poor layers. These are called banded-iron formation rocks. The dark layers indicate that iron reached the sea before reacting with oxygen. The light layers are thought to represent some type of seasonal fluctuation.

About 2 billion years ago, banded-iron deposits disappeared and red beds began to form. Red beds are red rock deposits whose color comes from iron oxidized by oxygen in the atmosphere. Red beds suggest that atmospheric oxygen was present in enough quantity to oxidize iron in the sediments on land. The reddish walls of the Grand Canyon and throughout the southwest of North America illustrate the pervasive oxidation of iron in sediments exposed to an oxygen-rich atmosphere. The transition to an oxygen-laden atmosphere had begun.

It is now some 2 billion years ago. Early marine algae and bacteria are multiplying and photosynthesizing, adding more and more oxygen to the atmosphere. However, environmental conditions at the surface are still quite dangerous to marine organisms. When ozone forms from the disassociation of oxygen molecules in the atmosphere, it shields the surface from harmful ultraviolet radiation. On early Earth, without plentiful oxygen in the atmosphere, there is no ozone shield to prevent organisms at the surface from literally frying in the sun. In addition, organisms that use oxygen to obtain energy do it by causing oxygen to react with organic matter, a process known as oxidation. But oxygen is so reactive that cells must evolve a means to harness this powerful source of energy without burning up in the process. The sun remains a relatively untapped source of energy for most of Earth's life-forms and their growth is limited.

At about 1 billion years ago, enough oxygen is in the atmosphere so that an effective ozone screen is present, and organisms have developed the means to use oxygen efficiently and safely. Surface waters are now a good place to live; the sun's energy can be used, and marine plants begin to flourish. The climate and ocean are slightly cooler, and large continental landmasses have formed.

Some 750 million years ago, our story's setting begins to change. Through the movement of tectonic plates over Earth's surface, what were once separate "rockbergs" have become one gigantic supercontinent straddling the equator, elongated in an east-west direction. Plate tectonics, the process responsible for the movement of the continents, the creation and destruction of ocean crust, and much of Earth's restless nature, begins early; it will have an extraordinary impact on how Earth, the oceans, and life evolve. Ancient rocks and glacial debris suggest that much of the supercontinent's surface is covered by ice and Earth may be in the grips of its first and coldest Ice Age ever; ice and snow cover the land even near the equator. According to some scientists, the planet at this time resembles a giant snowball, but this concept remains controversial. Researchers are unsure what could have caused such a great Ice Age; new theories focus on the influences of the continents being positioned around the equator. By some 590 million years ago, however, the planet has again warmed and the stage is set for the next act of our evolutionary drama.

It is now the end of the Precambrian and the beginning of the Paleozoic era, some 550 million years ago. Life in the ocean is proliferating. There has been a great evolutionary leap from very primitive life-forms to more advanced and abundant creatures. For years this puzzled geologists and they searched for a missing link in the fossil record. Then in 1946, geologist R. C. Sprigg discovered the impressions of strange soft-bodied creatures fossilized in the ancient beach sands of the Ediacara Hills in southern Australia. The most abundant type of fossils are a sort of circular impression. The imprint resembles modern jellyfish: hence, this time period's nickname, the Age of Jellyfish, just before the Paleozoic era, about 600 million years ago. Also preserved in the Ediacaran rocks are the tracks and burrows from wormlike animals, odd bottom-dwelling forms, and a strange frondlike organism. Many of the organisms in the Ediacaran fauna are difficult to place in terms of contemporary marine species. Some scientists believe that they are related to sea urchins (echinoderms), worms, and crustaceans (arthropods). German paleobiologist Adolf Seilacher offers an alternative interpretation of the fossilized creatures. He suggests that these odd-looking organisms are not related to modern species, but represent life-forms that became extinct when their vulnerable mattress-like bodies fell prey to newly evolved predators. Since their discovery, Ediacaran fauna have been found on every continent except Antarctica, but they seem to disappear from the fossil record before the beginning of the Paleozoic era. It is unclear whether the marine organisms of the Ediacara became extinct because of a catastrophic event or changing environmental conditions or they were simply eaten by more successful predators.

The Ediacara fauna dramatically illustrate sampling problems in the study of the ancient seas. For many years, geologists assumed that before the Paleozoic era, Earth was essentially devoid of life—not because evidence showed there was no life, but because we could not find evidence of life. Life that existed in the sea before the Paleozoic was essentially soft-bodied, having neither a skeleton nor a shell. Preservation as a fossil was therefore geologically miraculous. When most soft-bodied marine organisms die, they fall to the seafloor and rapidly decay. If, somehow, their remains are buried quickly by soft mud or sand, the chances for preservation greatly improve. And if the surrounding sediments are bathed by water rich in minerals, such as silica or calcite, rocks may form with the impression of the soft-bodied organism intact. Fossilization is much more likely if an organism has a shell or skeleton; this is one reason we know so much more about life in later time periods. Once a fossil is found, through sheer luck or clever deduction, the quality of preservation influences our ability to interpret what it was and how it lived. Furthermore, our knowledge of present-day species may influence how we interpret organisms that in fact may have looked nothing like creatures alive in the modern sea.

The Paleozoic Era: Life Blooms in the Sea

IT IS SOME 550 million years ago in the Paleozoic era, the time of ancient life, and our story's backdrop is about to change again. The gigantic supercontinent still lies mainly along the equator, but soon large rifts tear apart the huge landmass and the sea spills over, creating extensive shallow-water regions. Over the next 200 million years, the continents separate and drift poleward. Rocks and fossils suggest that ocean temperatures range from 20° to 40°C (68° to 104°F), its chemistry and salt content are fairly similar to the modern sea, and the atmosphere contains increasing levels of free oxygen. The vast, warm, shallow-water habitats of the sea are now the perfect setting for an explosion of life.

The beginning of the Paleozoic, known as the Cambrian period, marks a time of unparalleled biological innovation and strange diversity in the sea. Within a span of 10 to 30 million years, marine life blooms radically and the morphological plans for all creatures on Earth are unveiled. Consequently, this period has been dubbed the Cambrian explosion or the biologist's big bang. The ancestors of crustaceans, shellfish, sea urchins, sponges, corals, worms, and other organisms are born.

For the first time, organisms begin to use minerals such as silica, calcite, and calcium phosphate in seawater to create shells or skeletons. In other words, creatures develop hard parts: shells, spines, and scaly plates.

The very first fauna with hard parts are small-shelled creatures; some look similar to modern-day forms, and others appear as strange tiny blades, tubes, plates, and cups. In Stephen Jay Gould's enlightening book, *Wonderful Life: The Burgess Shale and the Nature of History*, he indicates that paleontologists, with honorable frankness and definite embarrassment, named these first puzzling creatures simply, "the small shelly fauna." Over time, the small shelly fauna disappear, but soon thereafter the most famous of Cambrian fauna arrive, the creepy-crawly trilobites (Figure 5). Those with a special affinity for the trilobites, have christened the Cambrian, Trilobite Time. Their abundance and greatly enhanced preservation due to a mineralized exoskeleton make trilobite fossils so common that later they will be bought for a reasonable price at most museum and nature stores.

Trilobites, named for their oval-shaped three-lobed bodies, dominate the seas for about the next 100 million years. The seafloor is infested with them. Many are small, less than 20 centimeters long, and some are large, over half a meter in length. Most trilobites crawl about the bottom scavenging food, a few swim, and all prey upon the unwary. The horseshoe crab is somewhat reminiscent of its predatory trilobite ancestors.

Along with the trilobites come a number of other crustaceans, the clamlike brachiopods, echinoderms, and an odd group of cone-shaped sponges with a hard, calcareous skeleton. The brachiopods are filter-feeding, shelled organisms resembling clams that live anchored to the seafloor by a fleshy stalk or spine or simply rest in place. Echinoderms are named for their spiny skin; they include sea urchins, starfish, and a stalked flower-like organism called a crinoid, or sea lily. They are headless creatures, knowing no forward or backward, and all modern forms have

Figure 5 Fossil trilobite.
(Courtesy of the Smithsonian Institution.)

a five-sided symmetry. The Cambrian ocean swarms with crawlers, burrowers, a few swimmers, some floaters and those attached to the seafloor. Corals begin to grow and form primitive reefs, while jellyfish float in pulsating waves above (Color Plate 2a). Although many of the life-forms that developed during the Cambrian explosion are the ancestors of modern marine organisms, some scientists believe that others were strange creatures whose likes will never again be seen in the sea.

Fossils of the Cambrian period have now been found throughout the world. The first, most famous and controversial window into the Cambrian ocean was provided by the rocks of the Burgess Shale, an outcrop in the Canadian Rockies of southern British Columbia. In 1909, Charles D. Walcott, secretary of the Smithsonian Institution, discovered the first Burgess Shale fossils. Aided by family members, he spent years excavating fossils from the dark rock layers of the Burgess Shale and eventually supplied over 65,000 specimens to the Smithsonian's National Museum of Natural History (Figure 6). As later subsequent studies would indicate, the animals of the Burgess Shale had lived on a gigantic coral reef at the edge of a towering limestone cliff. Then, in one swift and muddy instant, they were killed, buried by a powerful underwater avalanche. The resulting fossils contained not only evidence of the first organisms with hard parts but also a diverse and abundant array of soft-bodied creatures—a paleontologist's dream. Walcott ultimately identified 100 of the 170 species now recognized. Some scientists criticize his interpretations because he attempted to classify the ancient creatures based on the body plans of organisms in the modern ocean. However, Walcott's enormous contribution to our understanding of the ancient seas remains uncontested.

For several decades after Walcott's work, the Burgess Shale fossils received little scientific attention. Then, in the late 1960s, a group in Cambridge, England, led by paleontologist Harry Whittington and his two students, Derek Briggs and Simon Conway Morris, began an extensive reexamination of the Walcott quarry and the Burgess Shale fossils that lay in storage. Using powerful microscopes they meticulously studied the Burgess Shale fossils. They also used a dentist's drill to unveil surfaces that had been hidden for years by hardened sediment. As their work progressed, the ink-black shale began to tell of creatures never before seen. In Gould's wonderful account of the Burgess Shale fossils, he emphasizes the peculiar nature of these "weird wonders."

Much controversy surrounds the Burgess Shale fossils, what they mean with regard to evolution, and if they represent ancestors of modern life-forms or unsuccessful blueprints that led to extinction. Walcott originally described one animal as a polychaete annelid—a segmented worm. Later, Simon Conway Morris found spines along its trunk that were unlike any polychaete, and in recognition of its "bizarre and dream-

Figure 6 Walcott and others excavating fossils from the Burgess Shale in British Columbia.
(Courtesy of the Smithsonian Institution.)

like quality" he named the animal, *Hallucigenia* (Briggs et al., 1994). In 1977, Morris depicted Hallucigenia as a wormlike creature that walked on stiltlike spines with seven waving tentacles on its back. In the early 1990s, scientists studying well-preserved specimens from China suggested a different interpretation. They proposed that Morris's *Hallucigenia* was in fact upside down. It was really a caterpillar-like creature with dorsal spines for protection and long, tentacle-limbs for crawling. Furthermore, what Morris had identified as the back was in fact the front, and front the back. Morris and others now believe that *Hallucigenia* is an ancestor of modern arthropods, a group that includes organisms such as crabs, spiders, and insects. Another of the Burgess Shale animals is *Anomalocaris*, or "odd shrimp." It is the largest—up to half a meter in length—and possibly the most voracious. It has stalked eyes, a squidlike body, a circular mouth and toothy jaw, and giant limbs attached to the front of its head. Its powerful jaws and predatory nature have earned it the name Terror of Trilobites. *Anomalocaris* was originally thought to be unrelated to modern-day species, but some now believe it may be another early relative of the arthropod group. Work and debate on the weird wonders of the Cambrian period from the Walcott quarry and elsewhere continues.

Returning to our story in the early Paleozoic sea, the characters have now become a diverse array of strange and familiar marine creatures. Weird spined and shelled organisms roam about, while scattered on the seafloor lie horn- and saucer-shaped corals, cone-shaped sponge, and numerous clamlike brachiopods. Bryozoans, often called the "moss animals," blanket the rocks and rubble with colorful calcium carbonate secretions; orange, purple, and green colonies helping to form growing piles of limestone. Small, simple bean-shaped crustaceans called ostracods lie in the sediments alongside the fusilinids, flat tubular-shaped foraminifera. The foraminifera are single-celled, amoeba-like animals that live within small calcareous shells. In the Paleozoic, they all live on the seafloor, but later, new species will rise to a life of floating within the water. The small shells of foraminifera preserved in the ocean's sediments will eventually prove invaluable to scientists studying Earth's past. In the sea's open waters tiny floating animals and plants, the plankton, are also becoming more numerous and diverse. The radiolarians, a group of small floating animals with beautifully constructed shells of silica, have arrived, and nearly all the major invertebrate groups with preservable hard parts are now present.

What could possibly have caused this rapid blooming of life in the Cambrian ocean? Several theories have been offered. One hypothesis relies on Darwinian evolution, asserting that life diversified because of competition between species and natural selection: a classic case of survival of the fittest. Only those organisms that evolved protective armor or hard parts could withstand the onslaught of new and hungry predators. However, the rapid pace of evolution within this period seems at odds with Darwin's concept of a more gradual process of natural selection. Another hypothesis is that environmental change facilitated life's bloom in the Cambrian. The earlier supercontinent had split up, and the distribution of land and sea had changed. Increased levels of oxygen and warming may have made the new seas more conducive to life, or perhaps the movement of the continents was such that oceanic regions were altered and enlarged, and more habitats for specialized lifestyles were created. Others speculate that a catastrophic asteroid impact occurred, changing oceanic conditions, and clearing the way for life's bloom. And finally, there are those who suggest that sex is the real answer to biology's big bang. Prior to the advent of sexual reproduction, organisms replicated essentially by self-cloning. This was a poor way to mix genetic material and allow for change. With sexual reproduction, genetic material could be rapidly mixed within a species and more quickly allow for mutations to evolve into new life-forms. But little evidence supports any one of these theories; most assuredly the quest to answer "Why then?" will continue.

It is now 450 million years ago. The ancient seas are infested not only with the creepy-crawly trilobites but with larger, more efficient predators. The squidlike cephalopods and fish are beginning to emerge. The first fish are jawless, eel-like creatures that swim along the bottom and scoop or slurp up mouthfuls of water and organic-rich mud. Modern representatives of these early fish include the lamprey and the hagfish (Figure 7), more appropriately known as the slime fish for the excessive quantities of slime they exude upon capture. Many of the early fish are small and have a "shell skin," an armor of bony scales and plates. The weight of the armor keeps them near the bottom, and their lack of fins limits their maneuverability. Later, fish will evolve jaws and become excellent browsers and

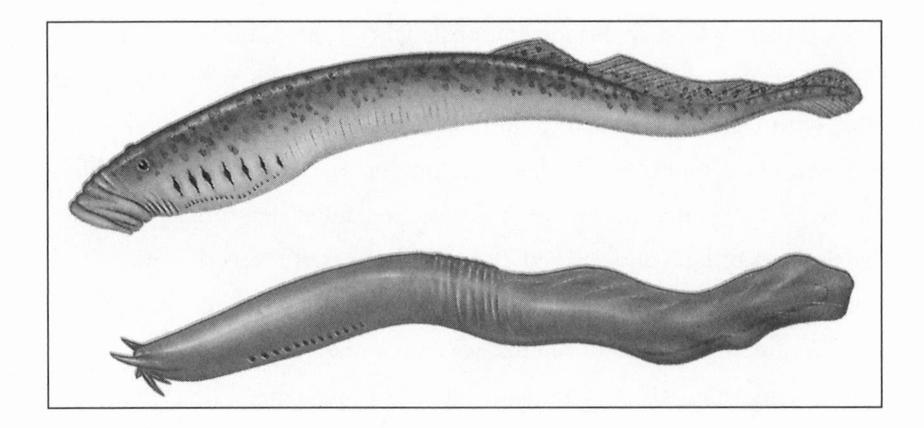

Figure 7 Sketches of a sea lamprey and a hagfish; actual lengths are about 60 centimeters.

(Reprinted by permission of John Wiley & Sons, Inc. from Exploring Ocean Science, *Keith Stowe, © 1996, John Wiley & Sons, Inc.)*

more efficient predators. Some will stay small, with sharp spines radiating out in all directions, while others will grow in size. One of the earliest jawed fish is a monster by any standards, reaching 9 meters in length, with huge jaws that could open exceptionally wide to engulf prey (Figure 8). The coelacanth is another primitive jawed fish. Once thought to have gone extinct, a coelacanth was caught in the net of a fishing trawler in the 1930s, south of Madagascar. More recently, coelacanths have been caught in the deep waters off Indonesia and in the Indian Ocean. Pale blue, with silvery markings and large platelike scales, the coelacanth is called a living fossil, unchanged for some 400 million years. Several scientists have rather doggedly, but unsuccessfully, pursued the collection and study of live coelacanths.

Some believe that at about this time, around 440 million years ago, Earth goes through a severe climate change. Glacial debris found in North Africa, Brazil, and Arabia suggests that the planet again plunges into an ice age. Water from the sea becomes trapped in expanding ice

Figure 8 *Fossil of a large primitive fish with a huge mouth and platelike armor. (Courtesy of the Smithsonian Institution.)*

sheets, causing a dramatic fall in sea level and the draining of many shallow marine habitats. Some organisms are totally wiped out, while others survive and recover—a pattern that is to be repeated throughout Earth's history. Life evolves, diversifies, and then is periodically blasted by an event so destructive that a biologic catastrophe ensues. Some life-forms are decimated; others survive, recover, multiply, and evolve.

It is now some 400 million years ago in the early Devonian period. The basic jawed-fish design is in place and they dominate the seas. One strange Devonian species has the back end of a modern fish with a sharklike tail and a primitive front end with a bony box for a body, legs like a crab, and eyes and mouth like a frog. Over time, the heavy armor of the early fish is completely lost and replaced by lighter, more flexible scales. Interestingly, teeth in sharks and all other vertebrates, including humans, are believed to have derived from the scales or armor plates of the early fishes.

For millions of years invertebrates such as trilobites and echinoderms have ruled the sea, but their mobility is limited by an exterior skeleton. When the fish evolve with an internal skeleton of bone, they gain the advantage. An internal skeleton allows the fish to grow larger and more muscular. They are better swimmers and more mobile, more streamlined, faster. Their paired fins allow them to brake, turn, and maneuver; as on an aircraft or submarine, they also prevent roll, pitch, and yaw.

Sharks, cartilaginous fish, also originate in the Paleozoic era. The first sharks are similar to modern varieties, with an elongated body, large triangular fins, and a sharply upturned tail with a lower lobe. Early sharks evolve two dorsal fins and a pair of pectoral fins, and some have a bony spine or brushlike structure on the front dorsal fin. Each shark has a jaw lined by a conveyor belt layer of gum tissue. Within the shark's revolving gums, sharp calcium phosphate teeth are continually created and can be replaced every few days or weeks depending on the species. Because sharks and rays lack true bones, their ancient ancestors are known mainly from fossilized teeth, or in rare instances from the preserved microscopic teeth that make up their skin. Fossil shark teeth are relatively abundant—because in its lifetime, one shark can produce thousands of teeth.

The bony fishes and the sharks spread and diversify. Members of the early fish clan can even move about on land and survive in wet pools of mud. The primitive lungfish, named for its accessory lung, can breath air and waddle about the land using a pair of sturdy, muscular fins. It is easy to imagine how the lungfish may have later evolved into a land-dwelling creature.

At about this time, some 400 million years ago, Earth's surface begins to look very different. Plant life emerges from the underwater world onto land, the first insects take to the sky, and animals begin to roam the shores. Moss and ferns carpet the previously barren landscape in a new shade of green, and strange forests take root. Great swamps replace some of the earlier marine habitats, and a dry wind blows over vast desert regions. Competition within the sea and along the shoreline becomes fierce and animals are forced to move onto land, seeking refuge and new sources of food. The first creatures to leave their wet world are the early amphibians, ancestors of modern frogs, toads, and salamanders. Fossils and preserved tracks suggest that they typically lay in streams and swamps munching on insects, fish, and each other, but every once in a while they venture out onto dry land. Amphibians can only spend a portion of their time out of the water, having to return to the sea to lay their eggs. The transformation to a land dweller is still incomplete. Those first forays from the sea to land must be a marine creature's nightmare—the powerful heat of the sun, the inevitable tug of gravity on the body, weird-looking food, and unknown predators. But somehow life finds a way, and animals evolve the skeletal and cellular requirements to adapt to life on land.

At about 300 million years ago, the first true land creatures appear—the reptiles. Unlike the amphibians, reptiles produce large eggs with a leathery or calcareous shell that can be laid on land; this removes their need to periodically return to the sea. The first reptiles are small salamander-like creatures, but they grow and diversify quickly. Some are herbivores, feeding on plants, while others become carnivores, meat eaters. Several of the big lizards have a large, fanlike "sail" on their back, possibly to help regulate their temperature: nature's original solar panel.

It is now 270 million years ago, near the end of the Paleozoic in a time called the Permian Period. The slate of characters is once again full, and the ocean is bustling with life. The climate has warmed and shallow seas cover much of the planet's surface. The marine environment is conducive to undersea growth. Filter-feeding organisms are now dominant creatures on the seabed. Tall, flower-like crinoids wave in the currents, some reaching heights of up to 3 meters. Crinoids, the lilies of the sea, have long, thin, cylindrical stalks made up of fitted limestone disks each topped by a cup of tentacles. Brachiopods by the hundreds sit quietly filtering water through their clamlike bodies. And the bryozoans spread over the rubble and build colorful colonies shaped like twigs, fans, and fingers. On reefs, corals take a back seat to other limestone-producing organisms. Sponges, bryozoans, algae, and foraminifera accumulate into massive limestone structures. These growing undersea hills of calcium carbonate teem with crawling, swimming, and stationary marine creatures. The bony fishes and cartilaginous sharks have spread and diversified. Even the swamps are full of eel-like sharks with huge jaws and double-fanged teeth. The seas are alive with life, huge reefs are crowded with marine creatures, the surface currents are thick with floating plankton, and fish are plentiful in Earth's waters.

But about 250 million years ago, something happens and life on Earth is almost annihilated. A mass extinction brings the curtain down on the Permian and Paleozoic. During the "great dying" 90 percent of all existing species are killed and a great biologic restructuring of Earth takes place. The trilobites and several species of shellfish and coral are completely wiped out. The small fusilinid foraminifera just disappear and the brachiopods, bryozoans, crinoids, and fish experience major losses to their ranks. On land, the number of amphibians and insects are dramatically reduced.

Many believe that the mass extinction at the end of the Permian occurred over a relatively short period of time, within several million years, but no one has found definitive evidence of a cause. Theories to explain this biologic devastation include a severe climate change, a dramatic lowering of sea level, an upwelling of anoxic (oxygen-poor) water, the appearance of a naturally toxic substance, or the removal of a few key species in the food chain. Recent research suggests that the Permian extinction actually occurred within less than half a million years; this has led some scientists to conclude that a colossal asteroid or comet hit

Earth. Others believe that the extinction at the end of the Permian is related to changes in the configuration of the continents and the intensification of deep-seated volcanic activity. In Siberia, volcanic lava from this time forms a pile some 3 kilometers thick and 2.5 million square kilometers wide (Lamb and Sington, 1998). This great outpouring of molten material is thought to reflect a period of intense volcanic activity on Earth. Massive amounts of dust, carbon dioxide, steam, and sulfur dioxide could have been ejected skyward. Water in the atmosphere may have combined with sulfur dioxide to form sulfuric acid. Torrential downpours could then have immersed the surface in a caustic bath of acid rain. But even though this is one of the most influential events in the history of Earth and life, we still do not know how or why it happened.

The Mesozoic Era: A Midlife Crisis

WE ENTER THE Mesozoic era, some 250 to 245 million years ago, and a new world order ascends in the sea and on land. The setting is also changing. Plate tectonics has again rearranged the continents into one huge landmass that we call Pangea (Figure 9a). The new supercontinent stretches almost from the south to the north poles and covers about 40 percent of Earth's surface. One gigantic world ocean, relatively deep and twice as wide as the modern Pacific, surrounds Pangea; it will be called Panthalassa. In Panthalassa, winds and other surface forces create two large circular patterns of water movement called gyres, one in the Northern Hemisphere and one in the Southern Hemisphere. Extreme differences in water temperature occur along the east and west shores of Pangea. Sea level is relatively low, there are fewer shallow-water habitats around the continental margins, and the climate is arid. Both shallow- and deep-ocean water temperatures are still warm, and their salinity and chemistry are fairly similar to the modern-day ocean. Variations in climate occur with the seasons and latitude, but no extensive glaciers or ice sheets have formed at the poles.

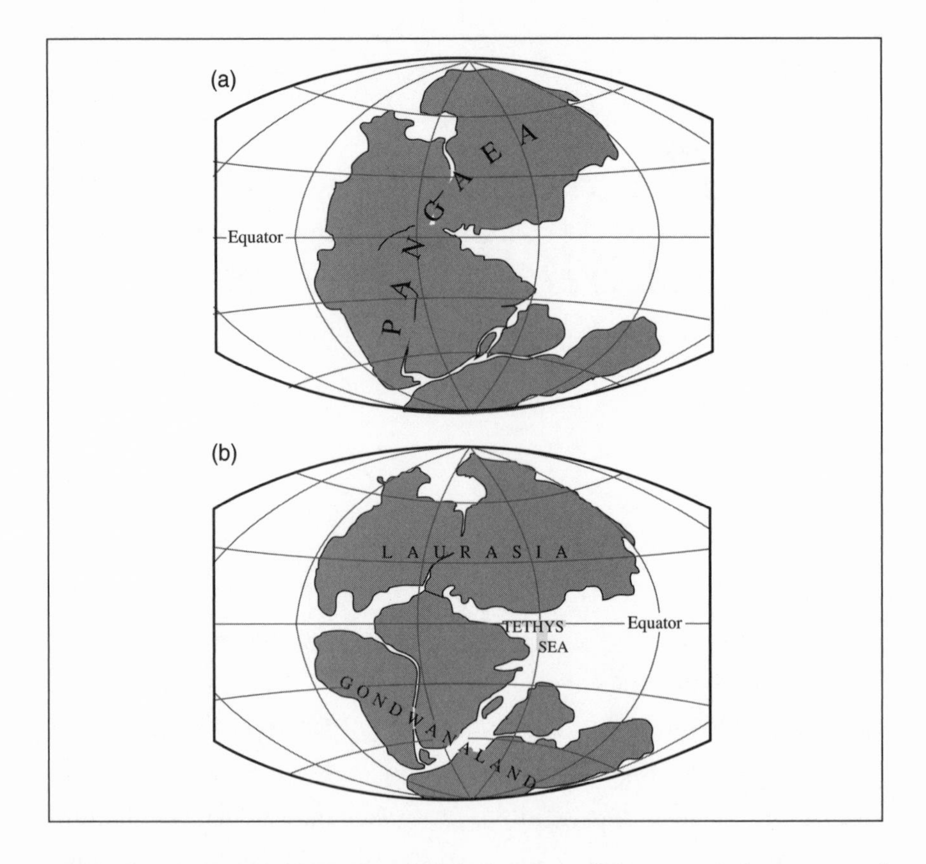

Figure 9 (a) *Pangea, 225 mya.* (b) *The breakup of Pangea and the creation of Laurasia and Gondwanaland, 200 mya.*
(Courtesy of the U.S. Geological Survey.)

Around 170 to 200 million years ago, the same processes that brought the continents together tear them apart and create two major landmasses, Laurasia in the north and Gondwanaland in the south (Figure 9b). In between, lying along the equator, a narrow seaway is formed, the Tethys. Flow within the Tethys seaway produces a great, globe-encircling current that transports heat throughout the huge world ocean. Soon the two continental landmasses splinter and the ancestral Atlantic and Indian oceans begin to form. Sea level rises precipitously, flooding the land and again forming extensive shallow marine environments. Earth's climate warms.

Organisms that survived the Permian-Triassic extinction are slow to recover, but within less than 100 million years both animals and plants are again thriving and some very big newcomers appear on the scene. On land, the reptiles have emerged from the Permian catastrophe, reduced in number but definitely not gone. Around 200 million years ago the reptiles give rise to the largest, most ferocious predators on Earth—the dinosaurs. Throughout the Mesozoic the dinosaurs proliferate, diversify, and become rulers of the land. Some of the dinosaurs take to the sea. The plesiosaur, with its short broad body and long neck, grows up to 3.6 meters (12 feet) long and uses large bony flippers for swimming. The ichthyosaur, looking similar to the modern porpoise, also has excellent swimming skills and becomes a superior ocean predator. And the mosasaur, a monster lizard equipped with a long body and flippers, joins the underwater fray. The ocean-going dinosaurs prey upon sharks and bony fish, and some use specialized teeth and jaws to munch on shellfish. Turtles also enter the sea during the Mesozoic and grow to gigantic proportions, almost 4 meters in length. The newly evolved birds become predators from the air, streaking like missiles into the sea. Fish evolve into larger, swifter swimming species. Sharks take on their contemporary appearance and diversify even more. Bite marks found on the fossilized bones of marine dinosaurs suggest that sharks are not only prey but predators of the sea-going reptiles. Rays evolve from their shark ancestors and begin to grace the oceans with their elegant underwater flight. Most are benthic, spending their time lying on the ocean floor, but some such as the manta ray will fly through the sea with unparalleled grace. Late in the era, skates, stingrays, sawfishes, and the first electric eels also appear.

During the Mesozoic era, the mollusks greatly diversify and expand in number. The word "mollusk" comes from the Latin term *mollis,* meaning soft and refering to a soft body living within a shell. Most of our modern-day shellfish—clams, scallops, oysters, snails—are mollusks. Some cephalopod mollusks have no shell at all. The cephalopods are a subgroup of mollusks that are smart, fast-moving, meat-eating hunters; they include squid, cuttlefish, octopus, nautiloids, and ammonites.

Cephalopods evolve in the Paleozoic and become some of the most abundant and successful marine animals of the Mesozoic. The nautiloids, a group of cephalopods from this period, are tentacled, soft-bodied organisms that live in a beautifully coiled shell. As the animal grows, it builds a newer, larger chamber inside its expanding shell, vacating the previous chamber and using it solely as a buoyancy control device, much like the apparatus of a submarine. By the time the creature reaches maturity, it has built a shell of dozens of chambers. Unlike snails, the coiled shell of the nautiloids lies flat in a single plane. Each chamber holds gas and fluid that counteract its weight, so that even as the organism is protected by a weighty hard shell, it is also a fast, predatory swimmer. The nautiloids are common in the Paleozoic and present during the Mesozoic; however, the only modern version is the chambered nautilus found in the deep South Pacific and Indian oceans.

The nautiloids give rise to the ammonites, organisms with thick, tightly coiled, ridged shells that can grow to gigantic proportions, spanning nearly 2 meters across (Figure 10). Like the nautiloids, they are swift and clever predators, moving through the water by forcefully expelling water through a tubular funnel, a sort of internal jet-propulsion system. Now, in addition to evading dinosaurs and sharks and diving bird torpedoes, marine creatures have to outmaneuver the armored, jet-propelled ammonites. Add to the mix marauding squid 20 meters long, giant 3-foot-wide clams, and crinoids towering 18 meters high and the Mesozoic ocean is a wondrous, but perilous place to live (Color Plates 2b and c). Along the shore it is no better, as large, speedy predatory reptiles, including the newly evolved crocodiles, lie in wait.

Figure 10 *Fossil ammonite shell.*
(Courtesy of the Smithsonian Institution.)

Life is also flourishing near the ocean's surface. The small floating organisms, the plankton, are multiplying and branching out. The foraminifera have added planktonic forms to their ranks, and the calcareous coccolithophores have arrived. These are small, planktonic spheres of golden-brown algae whose surface is covered by tiny, round shieldlike disks of calcium carbonate called coccoliths (Figure 11). By the end of the Mesozoic, in the Cretaceous period (140–65 mya), the coccolithophores are so abundant that they produce huge accumulations of fine, white ooze (mud) on the seafloor. Over time, these sediments compact and harden into a thick deposit of fine-grained, white limestone; called chalk. The famous white cliffs of Dover are made of millions of tiny coccoliths deposited during the Mesozoic.

Another plankton group appearing on the scene is the dinoflagellates, small unicellular algae with a filmy cellulose covering, similar to that of

Figure 11 Photograph of a coccolithophore under a scanning electron microscope. Scale bar is 5 microns in length.
(Courtesy of Vita Pariente, Texas A&M University.)

pollen, and two whip-like tails, or flagellae, for propulsion. The dinoflagellates will grow to become an important component of the plankton. In the modern ocean, they produce the unsightly and toxic phenomenon known as red tide, as well as much of the ocean's startling nighttime bioluminescence.

Late in the Mesozoic, the diatoms also arrive and become another significant group of phytoplankton. The diatoms have small pillbox-shaped shells of silica and can be single individuals or form long, delicate chains. In the modern sea, diatoms are especially abundant in cold-water regions, but in the ancient oceans they are common in areas of volcanic activity. Volcanic ash and dust provide the silica that diatoms need to construct their tiny shells. The siliceous radiolaria are also growing in abundance and the pteropods, or sea butterflies, make their entrance into the sea. Pteropods are delicate creatures, an upside-down swimming snail with a foot that has been modified to form butterfly-like wings. Some have a small, paper-thin, tubular shell of aragonite, a form of calcium carbonate; others exist naked, free of their tiny shells.

On the seafloor, algae are also diversifying. Marine plants that have a skeleton of calcium carbonate, the calcareous algae, expand and evolve into new species. One such algae, called *Halimeda*, has an internal skeleton made of articulated plates. Each plate resembles a bleached flake, similar to a piece of oatmeal or a cornflake. When the *Halimeda* plant dies, these skeletal plates disarticulate and fall to the seafloor, producing carbonate sediment. Halimeda will become one of the most important producers of sand in tropical environments.

When organisms die or are washed into the sea, they fall to the ocean floor, and if oxygen is present, they begin to decompose. However, if little oxygen is available and organic matter is abundant, organisms may get buried without decomposing. Organic matter buried deep in the earth can be transformed, where pressure and temperature increase, into precious hydrocarbon products, such as oil and gas. Many of the great oil fields exploited today are products of the Mesozoic ocean and its abundance of tiny marine creatures. For such large amounts of organic matter to have been buried, the oceans of the Mesozoic must have periodically become oxygen-deficient. It is unclear why this happened or how long it lasted. One theory suggests that oxygen depletion occurred because of the ocean's sluggish flow. There was little temperature difference between surface and bottom waters in the Mesozoic ocean. It is thought that this made for slow circulation, slow renewal of oxygen in deeper waters, and the periodic deposition of large amounts of organic carbon on the seafloor. Oxygen may have been further depleted by sporadic influxes of very dense, salty water created by evaporation in restricted shallow basins. Near the end of the Mesozoic, in the late Cretaceous period, the deposition of these organic-rich, anoxic sediments becomes rare. The continued opening of the Atlantic and Indian Oceans or intermittent episodes of climatic cooling may have sufficiently increased ocean circulation.

The climate in the Cretaceous is generally warm, sea level is high, and shallow seas cover much of the planet. Periodic changes in sea level, climate, and ocean mixing have occurred, but on the whole it is a warm and equable time in the sea—and it is about to get warmer. The rate of seafloor spreading is rapid, possibly the fastest ever, and volcanic activity on Earth is rampant. Beneath the floor of the western Pacific Ocean (the remains of Panthalassa), a plume of hot molten rock wells upward from deep inside Earth and creates intense volcanic eruptions. Hot lava and gas burst forth in such magnitude that sea-level rises and the climate warms even more.

Within the sea, however, abundance and diversity endure. Large predators, swift swimming fish, and small floating plankton abound. Corals have now rejoined the other major reef builders and coral reefs extend into higher latitudes, fringing the continents and topping the peaks of submerged, extinct volcanoes. Oysters and a coral-like bivalve called a rudist grow in great abundance. In quiet lagoons, sea urchins, starfish, brittlestars, and a bevy of other echinoderms are thriving (Color Plate 3). So many calcium carbonate–producing organisms grow within the ocean during the Cretaceous that the deposition of their skeletons and shells create some of the most spectacular limestone deposits ever. But life's profusion will not last.

At the end of the Cretaceous period, some 65 million years ago, it happened again—another great biologic catastrophe. This is often referred to as the K-T boundary, signifying the end of the Cretaceous period (K in geologic mapping terminology) and the beginning of the Tertiary period. Within a few million years, an instant in geologic time, the entire dinosaur lineage, large and small, is completely wiped out. Some speculate that on land, no species weighing over 55 pounds survives the crisis. About 75 percent of all existing animals and plants disappear, and those living in the ocean, particularly near the sea surface, are especially hard hit. The ammonites and other mollusks, some of the hard corals, sea urchins, fish, and several marine reptiles all go extinct. The planktonic foraminifera and coccolithophores nearly disappear, and many other benthic species are lost, particularly in the shallow regions.

For many years, scientists were puzzled by the mass extinction at the K-T boundary. Then in 1980, geophysists Luis Alvarez and his son Walter proposed a radical idea. They suggested that some 65 million years ago a massive asteroid hit Earth. The impact produced such an immense explosion and wrought such a severe climate change that most of life on Earth was annihilated. The basis of this theory was a thin layer of iridium-rich soil first found in Italy and then worldwide, all dated at the K-T boundary (Color Plate 4). Iridium is uncommon on Earth's surface, but Luis and Walter Alvarez observed that high levels of iridium are frequently found in meteorites. They speculated that if a large enough iridium-rich asteroid hit Earth and exploded, a tremendous number of iridium-rich particles could be ejected into the atmosphere and eventually settle as an iridium-rich blanket of dust covering Earth's surface. Calculations suggested that a meteorite would need to be some 10 kilometers (6 miles) across to have produced the amount of iridium present in the soil (Hartmann and Miller, 1991).

Additional evidence has since been discovered to support the impact of a gigantic meteorite at the K-T boundary. Fragments of altered quartz crystals, called shocked quartz, have been found in the K-T soil layer and are indicative of a powerful explosion. Tiny glass droplets also found immersed in the boundary sediments are consistent with an explosion in which hot molten rock sprays out and then very quickly cools and resolidifies. Scientists also found chaotic deposits of rock debris whose nature suggests the powerful bulldozing action of a towering tsunami. If it was at least partially in the ocean, the impact would have created a giant wave that swept over the surface and pushed sediments into a jumbled pile of rocky debris. The K-T layer was also found to contain high concentrations of soot, indicative of a great burning event. Researchers calculated that to obtain the amount of soot found, more forests would need to burn than exist on the entire planet today. Finally, in 1990, on the north coast of the Yucatan Peninsula, geologist Alan Hildebrand discovered a partially submerged crater, some 180 to 200 kilometers across, associated with large amounts of shocked quartz and greatly fractured rock indicative of a massive impact. Now referred to as the Chicxulub Crater, it is so far, the best bet for the location of the cataclysmic K-T impact.

Many now accept that a colossal asteroid impact caused the biologic devastation of 65 million years ago. Modern space technology enables us to identify and track huge asteroids in space and observe meteorite impacts on other planets. In light of this knowledge and the size of other gigantic craters on Earth, the impact theory seems more plausible than ever. Nevertheless, some scientists are not convinced that an asteroid impact alone wiped out the dinosaurs; they feel that the mass extinction was a combination of cosmic and earthly events, including the intense volcanic activity of the time.

It is frightening indeed to imagine a giant asteroid hitting Earth. Upon impact, a tremendous explosion with the force of 100 trillion tons of dynamite occurs, ejecting solid rock debris, fiery molten lava, and dust high into the atmosphere, and maybe even space. The initial explosion and burning debris falling back to Earth create a global firestorm. The force of the impact triggers a huge tsunami that sweeps through the sea and across the land. And the shock from the impact transforms large quantities of nitrogen in the atmosphere into nitric oxide, producing caustic acid rain. If an organism survives the impact, the explosion, the fires, and the tsunami, it then receives a bath of blistering acid. As vast amounts of dust and ash settle in the atmosphere, day turns to night. Darkness blankets Earth for weeks, months, and maybe even a year or so. Without light for photosynthesis, phytoplankton—the floating plants of the sea and the base of the marine food web—are wiped out. With their major food source gone, the floating animals, the zooplankton, starve. In turn, the fish go hungry and finally, the larger creatures of the sea suffer. On land, a similar chain reaction occurs. The horrifying effects of a large asteroid impact evidenced at the K-T boundary have inspired many to investigate the probability of another catastrophic impact in the future, and to ponder the ultimate role of impacts in evolution.

The Cenozoic Era: Modern Life

THE END OF the Cretaceous period marks the beginning of the Cenozoic era, or time of recent life. It spans from 65 million years ago till the present and will be much different from the preceding eras. The blueprints for life have already been drawn, and now it is time to see who can adapt, compete, and withstand changing environmental conditions. Vulnerable or fragile species will have to contend with both predators and a varying environment; only the robust will survive. Huge mountain chains will be formed beneath the sea and on land, and forever alter the planet's climate. The oceans will see profound changes in temperature and circulation that impact the planet and its distribution of life. Mammals will become the dominant players on land and eventually give rise to our own ancestors.

We know vastly more about the Cenozoic era than we do about earlier times because we have access to a more complete fossil and rock record. Yet this period, about which so much has been written, represents a mere 1.5 percent of the planet's 4.5 billion-year history.

As the dust and ash from the K-T impact settle some 65 million years ago, Earth's surface begins to look familiar. There are large continents and widening ocean basins (Figure 12a). The early South Atlantic Ocean now lies between the African and South American continents. A narrow North Atlantic Ocean has just begun to form between North America and Europe. Australia, once attached to Antarctica, has broken away and is slowly moving northward, and India has separated from

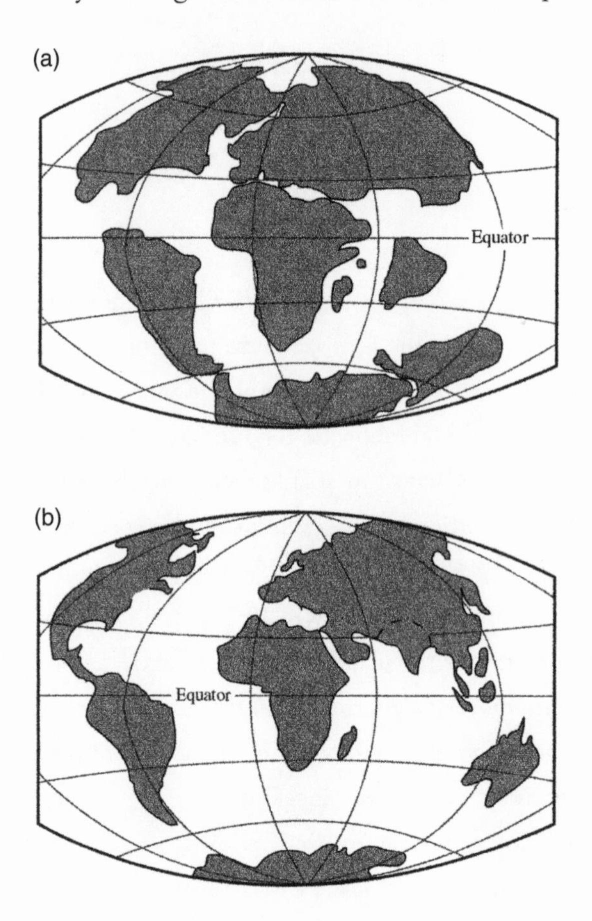

*Figure 12 Distribution of continents (*a*) as they appeared 65 million years ago and (*b*) as they are today.*
(Courtesy of the U.S. Geological Survey.)

Africa, moved north, and will soon collide with Asia. The changing position of the continents and opening of ocean basins in the early Cenozoic has a major impact on conditions within the ancient sea and, soon, on Earth as a whole. The Tethys Seaway and its equatorial circulation are being closed off. Australia and South America are moving northward, and with Antarctica nearing the South Pole, a new, nearly circular flow around the Southern Continent is developing. Earth is becoming a planet of multi-ocean basins and like a growing wound, a huge underwater rift and mountain chain is spreading across its surface. Eventually the mid-ocean ridge system will wind around Earth for over 60,000 kilometers and become the planet's largest, though submerged, mountain chain (Color Plates 5 and 6).

A recent theory suggests that early in the Cenozoic, 60 to 55 million years ago, when Earth and its creatures were recovering from the K-T impact, a period of strong global warming occurred both on land and beneath the waves. A huge volcanic blast in the Caribbean Sea and a megaburp of methane from under the sea floor may have occurred, causing even further warming on Earth. Some suggest that the warming of the oceans and a resulting stagnation of its deep waters caused another mass extinction of marine organisms, particularly many of the deep-sea foraminifera. On land, this episode of global warmth is thought to have opened the way for mammals to multiply and spread throughout the continents.

Those organisms that survived the great collision from space at the end of the Mesozoic are starting to recover, and as the Cenozoic progresses, a new world order begins. On the continents plants sprout and spread green over the land. The climate is warm and wet; tropical forests extend from the equator to the middle latitudes, and woodlands spread into the polar regions. Without the grazing of herbivorous dinosaurs, vegetation grows in lush abundance. Small burrowing mammals that survived the recent calamity expand across the continents and diversify, filling the void left by the dinosaurs. As they grow in size, they adapt to more specialized environments and spread into new habitats, such as the sea. Seals, sea lions, and walrus now frolic in the waves. Then the whales and porpoises appear.

The first whales are of the toothed variety: such as dolphins, killer whales (orcas), and sperm whales. Unlike their mammal cousins on land, they have no hair or fur; instead, they have a sleek skin to facilitate movement in the water. They evolve thick layers of blubber under their skin to keep them warm in the cold sea. The limbs that they used on land for walking are lost or modified for paddling. External nostrils move from the front of the head to the top and back of the skull, becoming a blowhole for breathing. The whales also have extremely flexible and collapsible lungs to facilitate deep diving. And the mammal ear is modified to sense underwater sound or vibrations, given whales, dolphins, and porpoises evolve the ability to echolocate. The baleen whales, such as the right whale and giant blue whale, evolve later; they have a fuzzy jaw, or baleen, that enables them to scoop up and filter vast quantities of krill, plankton, and small fish out of the sea.

The ancestry of whales has been a matter of debate for many years. While we do know that whales evolved from four-footed terrestrial mammals, we do not know which mammals those were, or why they abandoned the land to return to the ocean. Because of striking dental similarities, some paleontologists believe that whales evolved from wolf-like mammals. On the other hand, scientists using DNA data think that whales evolved from a group of animals that include hippopotamuses, camels, and pigs. Fossil whalebones some 50 million years old were recently uncovered in Pakistan, but rather than helping to solve the evolutionary debate, they deepen the puzzle; the bones do not support either theory (Wong, 1999), so the mystery goes on.

One reason the marine mammals are able to so successfully colonize the sea during the Cenozoic is that an exceptional abundance of food is available. With much of their competition gone, Mesozoic survivors proliferate in the ocean. The phytoplankton, including diatoms, dinoflagellates, and coccolithophores, multiply rapidly, and thus, so do the grazers, the filter feeders, and the scavengers. No new major groups form, but the crustaceans, radiolarians, shellfish, corals, bryozoans, sea urchins, starfish, and foraminifera blossom. One large coin-shaped foraminifera reproduces in such quantity that their calcium carbonate shells accumulate in thick deposits on the seafloor. Over time, this foraminifera-rich sediment hardens into limestone and is thrust upward during a major mountain-building event. Later, the Egyptians will recognize the beauty and utility of the foraminifera-rich limestone and quarry it to build the pyramids and the mysterious Sphinx.

The brachiopods decline in the Cenozoic, and only a limited number of species will survive the test of time. In contrast, the mollusks prosper in the new oceans and establish themselves as one of the most plentiful and diverse creatures on the planet. As the mollusks diversify, they form a myriad of beautifully crafted and colored shells. Scallops, clams, oysters, snails, squid, octopus, and mussels begin their subtle reign in the sea. Corals once again become the dominant players on reefs, while bryozoans, though present, move to the back seat. The fields of undersea lilies disappear and the crinoids are relegated to a somewhat cryptic existence in the nooks and crannies of the seafloor. The bony fishes explode in number and spread throughout the seas as never before. Perch, bass, snappers, seahorses, sailfish, barracuda, swordfish, tuna, and many other scaly varieties abound. The cartilaginous fishes, the sharks and rays, also flourish. Shark teeth as large as a human hand have been found and dated from the early Cenozoic (Figure 13). Based on the massive size of the teeth, scientists estimate that the shark, *Carcharodon megaloden,* grew as

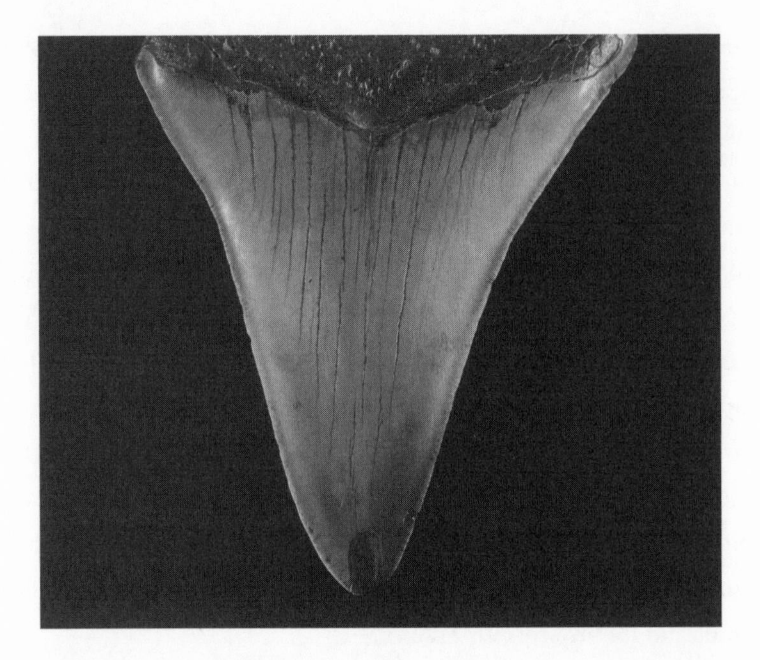

Figure 13 Fossilized shark tooth from an ancient ancestor of today's great white shark.
(Courtesy of the Smithsonian Institution.)

large as 12 meters (39 feet) in length, twice the size of its modern relative, the great white. Sea turtles, crocodiles, and sea birds also make it into the early Cenozoic ocean.

But soon the backdrop again begins to change. An episode of great mountain building occurs and launches a series of events that will result in the long-term cooling of Earth's climate. The North American plate shifts west, pushing into the Pacific plate, crumpling the edges of the North American continent upward. Sediments that once lay beneath the sea are heaped into towering, rugged piles, creating the Rocky Mountains. India slowly crashes into Asia and creates the huge Himalayan mountain chain. Here, the collision zone stretches for some 2900 kilometers, and mountain elevations reach up to 8854 meters above sea level. The Himalayas become the highest continental mountains in the world, and because plate convergence persists, they continue to rise at a rate of 1 centimeter per year. The Alps and the Pyrenees also form in the early Cenozoic. Organisms and sediments that once lay on the seafloor are thrust upward to form mountain peaks and valleys. Rain and wind pelt the new mountain cliffs. Weathering of recently exposed organic-rich, shallow-water sediments releases carbon dioxide into the atmosphere, causing the climate to warm. Erosion by wind and rain continues, and sediments wear off the mountain slopes, exposing rocks from deeper within Earth's crust. Because of their composition, these rocks do not release carbon dioxide into the atmosphere, but instead absorb it in great quantities, as do the incredibly dense stands of vegetation that now grow across the land. The concentration of carbon dioxide in the atmosphere decreases, and the global climate begins to cool. Latitudinal differences in temperature increase: the poles get much colder; the tropics cool but stay relatively warm.

In the oceans, water temperature also begins to fall and differences with latitude intensify circulation. Silica-producing plankton, like the diatoms and radiolarians, flourish in the cooling middle and low latitudes. Around 38 million years ago, about halfway through the Cenozoic era, an event in the ocean occurs that drastically alters the sea and indicates a major change in Earth's climate. The deep ocean fills with frigid water. Evidence from the shells of bottom-dwelling foraminifera suggests that bottom temperatures decrease by about 4 to 5°C (7-9°F) (Kennett, 1982).

Shells as ancient thermometers? Scientists have developed a clever technique for calculating past ocean temperatures from the oxygen isotope composition of foraminifera shells (Figure 14). Oxygen has two stable isotopes, oxygen 16 and oxygen 18. The ratio of these isotopes in a marine organism's calcium carbonate shell is dependent on the isotopic composition of the surrounding seawater and the water temperature during the organism's growth; in seawater, the ratio of oxygen isotopes is

Figure 14 Foraminifera shells picked from the sediment of a deep-sea core. (Courtesy of the Ocean Drilling Program.)

strongly controlled by the temperature of the water and the presence or absence of large amounts of ice on the planet. The isotopic composition of a foraminifer shell is measured with an instrument called a mass spectrometer. Then based on its oxygen isotope ratio and an estimate of the composition of seawater when the organism grew, scientists can calculate an approximate water temperature in the ancient ocean. The oxygen isotope method, particularly using foraminifer shells from deep-sea cores, has become one of the most important techniques used to study ancient ocean temperatures, sea level change, and Earth's past glacial history.

By around 38 million years ago a huge influx of cold water enters the deep sea. Many of the organisms living at or near the seafloor are killed by the sudden change in temperature. Robust forms or those that can migrate to warmer waters survive. But where has this cold water come from?

At about the same time that the deep sea is plunged into coldness, conditions around Antarctica become frigid. Australia has moved northward and cool water has begun to flow from the southern Indian and Pacific Oceans into a small embayment alongside Antarctica, the Ross Sea. This triggers the first large-scale formation of sea ice. When sea ice forms, freshwater is preferentially frozen, salt is excluded, and the surrounding seawater becomes saltier. Salty, cold seawater is very dense and sinks relative to surrounding, less dense water. The formation of sea ice around Antarctica thus leads to a cascade of cold, salty, very dense seawater downward into the ocean's depths, filling the deep sea with frigid water. Following this event, an ocean that is warm on top and cold on the bottom becomes the norm and patterns of flow driven by changes in temperature and salinity begin.

Some 20 million years ago, Earth is still changing. A rupture in the African continent widens, and the narrow Red Sea takes shape. The split causes Saudi Arabia to move north against Asia and seals off the remaining portion of the Tethys Seaway. What is left of the Tethys forms a nearly enclosed sea, the Mediterranean. The global transport of warm water around the equator is shut down for good. From now on, warm water circulation at the equator occurs only within individual ocean basins. In the south, South America has moved away from Antarctica, opening what is known as the Drake Passage. A true circumpolar current around Antarctica is formed, and a complete reorganization of ocean circulation in the Southern Hemisphere follows. A zone of fertile growth is formed in the far southern latitudes; fossils from New Zealand suggest that at about this time, an abundance of whales and penguins inhabit the area.

The oceans, continents, and atmosphere have nearly obtained their modern configuration (Figure 12b). The North Atlantic is widening and Greenland has separated from Europe. The Iceland ridge that separates North Atlantic waters from the Norwegian Sea begins to subside and eventually sinks below sea level. Cold, dense water forms at high northern latitudes and flows southward, mixing and eventually cascading down into the deep waters of the North Atlantic. Warm surface waters begin to flow northward along the western portion of the North Atlantic, creating the Gulf Stream. By this time, the atmosphere contains about 21 percent oxygen, 78 percent nitrogen, and less than 1 percent carbon dioxide and other gases.

The changes in the north and the isolation of Antarctica by a globe-encircling flow in the south lead to the development of large-scale ice sheets on and around Antarctica. Dark-colored surfaces, like the oceans, warm by absorbing much of the incoming radiant energy, whereas light-colored regions, like those covered by ice and snow, effectively reflect the incoming solar energy and stay cold. Greater ice cover increases Earth's reflectivity and causes further planetary cooling. A new pattern of circulation and temperature distribution develops in the sea. Temperature differences exist between the surface waters and the deep sea, and from the high-latitude poles to the low-latitude tropics. Marine organisms respond to these thermal gradients and gradually establish new biogeographic zonations. Warm-water species move to the surface and tropical waters, and cold-water flora and fauna go toward the poles and into the deep sea. The formation of an ice sheet on Antarctica has a long-lasting effect on the planet; it will keep Earth's climate from reaching the warmth of some 20 million years ago.

With global cooling comes a drier climate, and extensive savannas and woodlands begin to replace tropical forests. Grazing mammals evolve to make use of the growing supply of edible vegetation, and primates appear whose lineage will give rise to humans. Wind speeds increase, and along the coast, deep, cold marine waters well upward, bringing a wealth of nutrients to the surface. In response, phytoplankton bloom, zooplankton multiply, fish proliferate, and the entire food web prospers. The Gulf Stream becomes more powerful, driving warm water north, where it is chilled and returns southward below. Increased glacial ice on Antarctica locks up huge quantities of water causing a major lowering of sea level.

It is now some 6 million years ago. Cooling and the formation of ice sheets has lowered sea level some 40 meters (130 feet). The Mediterranean Sea is cut off from the Atlantic Ocean. Ocean water no longer pours in, and the Mediterranean becomes a series of large inland lakes. Sitting along the equator, the lakes are baked by the tropical sun and heat. As water evaporates, the lakes get saltier and saltier, like the modern Great Salt Lake. Sediment and rock cores drilled from deep beneath the Mediterranean Sea contain thick layers of minerals created by evaporation of seawater, such as gypsum, halite, and other salts, and indicate that the lakes periodically dry up and then refill. The periodic drying of the Mediterranean also has an important influence on the oceans, lowering its salt content and allowing sea ice to form at slightly warmer temperatures. More ice formation again increases the reflectivity of Earth's surface and causes greater cooling. Then somehow, global cooling is temporarily put on hold, and Earth begins to warm again. The sea rises, begins to fill the Mediterranean, and one of nature's greatest wonders is formed. Seawater pours in over the Straits of Gibraltar and encounters a 2000-meter drop from the top of the entrance to the bottom of the Mediterranean basin. A spectacular waterfall, or series of waterfalls, is created; water spills in with a force thousands of times greater than the flow of Niagara Falls (Hartmann and Miller, 1991). By 5 million years ago, the Mediterranean is once again full.

By now, there are extensive ice sheets in the Southern Hemisphere, but still none in the Northern Hemisphere. However, around 4 million years ago, Alaska connects with eastern Siberia, closing off the Arctic Basin, and pack ice begins to form in the Arctic Ocean. At this time, hominids, erect primates walking on two feet, begin to roam Africa. It is not completely clear why it took so long for ice to form in the Northern Hemisphere. Some believe that it was related to the final closing of the Central American isthmus between North and South America.

About 3 million years ago, great tectonic activity in the Caribbean causes the Isthmus of Panama to lift upward and close off the opening between North and South America. As a result, the Gulf Stream intensifies, more warm water flows northward, and precipitation at high latitudes is significantly increased. The barrier created by the land connection between the Americas also prevents marine species from migrating between the Pacific and Atlantic Oceans. As individual ocean basins become increasingly isolated, each begins to evolve its own distinct line of organisms. Because the Pacific is older, it contains a greater diversity of marine creatures than the Atlantic.

From now on, Earth's climate will seesaw up and down as the planet cycles through periods of extensive glacial cold and interglacial warmth. During the chilly glacial periods, cold air and water temperatures prevail, extensive land areas are covered with thick piles of ice and snow, and sea level drops. Rock debris and scour marks from glaciers indicate that ice often blankets much of the Northern Hemisphere. At the peak of glacial periods, global temperatures are some 5°C (9°F) less than contemporary temperatures, 2-mile-thick ice sheets blanket a third of Earth's land surfaces, and icebergs may cover half of the world's oceans. With sea level extremely low, more land is exposed and shorelines migrate up to 16 kilometers seaward. Rivers cut deep valleys and gorges along the exposed continental margins and the tropical region shrinks to just near the equator. Organisms in the sea and on land migrate to areas of sufficient warmth.

During interglacial periods, the air and sea warm. Sea level rises as snow and ice melt, and seawater expands as it warms. Land is submerged, shorelines move landward, estuaries fill with sediment, and glaciers retreat. The ice sheets melt from the land causing it to literally bounce back: the huge weight of glacial ice depresses Earth's crust; when ice is removed, the crust slowly rebounds. Today, regions in the Northern Hemisphere once covered by glacial ice are still rising at a rate of a few centimeters each year. Warmth-loving marine and terrestrial creatures spread north and south into higher latitudes as the air and sea temperatures rise. The zone of tropical heat expands as colder regions get smaller. Coral reefs proliferate and deposit thick piles of limestone. During one period of warmth and high sea level, some 125,000 years ago, a long, crescent-shaped coral reef forms just south of Florida. Later, during the next glacial period when sea level drops, this westward arching reef forms a series of islands, now known as the Florida Keys.

In the late Cenozoic, the global thermometer seesaws through as many as 30 major fluctuations and many more smaller variations. Ocean circulation, sea level, climate, and the distribution of life on Earth oscillate with the global glaciations.

A group of researchers recently simulated the global climate when glaciers last extended over the land some 18,000 years ago. They used all available data including that from ice and deep-sea sediment cores, oxygen isotope ratios in the shells of foraminifera, and computer modelling techniques. Their results suggest that thick ice sheets covered much of the Northern Hemisphere and were extensive around Antarctica, sea level was some 85 meters (about 275 feet) lower than today, and many forested areas became grasslands and deserts. The ocean was cooler overall, possibly by 2 to 3°C, and circulation was more energetic than during warmer periods (Hansen et al., 1993).

For decades, scientists have tried to explain the ups and downs of the planet's climate near the end of the Cenozoic. Today, the most widely accepted theory is that proposed by astronomer Milutin Milankovitch, who speculated that variations in Earth's orbit and rotation could cause periodic variations in the amount of solar radiation received by Earth, resulting in the temperature swings of the Cenozoic. Other factors, such as the changing configuration of the continents and ocean currents, mountain building, Earth's varying reflectivity, and the amount of carbon dioxide in the atmosphere, are also thought to play a role in controlling Earth's changing climate.

We reenter our story at the end of the last glacial period, some 15,000 years ago. As the climate warms, Earth emerges from its most recent deep-freeze. Ocean temperatures are on the increase and sea level is rising at a rate of about 8 millimeters per year. Circulation in the sea is slowing and warmth-seeking organisms again broaden their range. At about 5000 years ago the rate of sea-level rise slows to about 1–2 millimeters per year and the planet continues to warm. In the current century, Earth's warming continues—some would say, at an unnaturally accelerated rate. The atmosphere still contains about 21 percent oxygen and 78 percent nitrogen, but the amount of carbon dioxide is rising and heating up the planet. One of the greatest environmental controversies of our time—and possibly the most dramatic impact of human beings on the planet—is global warming. The year 1998 was the warmest year on record; record warmth is thought to have spurred the devastating El Niño of 1997 and the intense hurricane season of 1998. Will Earth continue to warm, and the sea level to rise, or as it has before, will the current warmth switch to cold, and the planet once again be thrown into a deep freeze? Time will tell.

As we come to the end of the story, we find that Earth and the oceans are still changing. Plate tectonics continue to move continents, raise mountains, and expand or shrink the ocean basins. The Atlantic Ocean grows while the Pacific, a remnant of the great world ocean Panthalassa, shrinks. The Mediterranean is getting smaller, while a new rift has split the African continent and will form a new ocean basin. Volcanoes continue to erupt both on land and under the sea, and islands are built and worn away. A great variety of marine creatures, born millions of years ago, continue to thrive in the sea. Some, based on the blueprints set forth in the Paleozoic era, have multiplied. Others have disappeared over the last few million years. Human influence may alter evolution's path. Pollution, overfishing, the introduction of exotic species, destructive fishing techniques, coastal development, and other activities threaten the health and well-being of the oceans and marine life. Organisms that have survived several cataclysmic events in the past may soon be wiped off the face of the planet by human indulgence and negligence. Life's history on Earth seems to include a series of long periods of gradual evolution, punctuated by events of rapid decimation. Rather than an incoming asteroid or meteorite, will humans be the instigators of the next biologic catastrophe?

This version of Earth's evolutionary epic has ended, but as more fossils are found and new technology is developed, much will be learned about the planet and life's origins. Undoubtedly, a great wealth of knowledge regarding evolution will also be unlocked through genetic and molecular research. The ancient seas are now the past and we enter the world of modern oceanography. Great disasters, adventures, and character changes are not gone with the past, but will inevitably be a part of the future, and advances in technology, exploration, and scientific research offer us a greater understanding of the ocean than ever before.

The Seas of Today

More has been learned about the ocean in the last half century than during all of preceding human history—yet the greatest era of exploration lies ahead.

—*Sylvia Earle*

The Science of the Sea Begins

*T*HE TIME IS today. As we enter the 21st century, the oceans, the product of a long evolution, cover some 72 percent of Earth's surface. On the surface, warm waters flow in great circular gyrations and sea level rises and falls with the rhythm of the tides and the undulations of the waves. In the deep sea below, the cold ocean moves in a slow and relatively steady course. The saltiness of the sea varies from site to site, but the proportion of its constituents stays amazingly constant—and similar to that which flows in our very own veins. Life teems in the ocean, in the warm, comfortable waters of the tropics, in the frigid pools of the poles, and even at its greatest depths.

For centuries the ocean was a mysterious and foreboding place. Before early explorers ventured across unknown waters, many feared the restless nature of the sea. Some believed that terrifying monsters lurked in its dark, boiling waters, while others thought that the sea contained great fields of fire, swirling whirlpools, or a narrow ledge that led to the very edge of Earth. The careless adventurer could easily fall off the land and be swallowed up into the deep, dark abyss. However, by the end of the 17th century, distant lands had been discovered and trade routes were established across the sea. Whaling and other fisheries began to foster a new interest in the ocean and the organisms that lived within its murky waters. In 1769, Benjamin Franklin published the first chart of the Gulf Stream (Figure 15). He used the length of time it took mail ships to cover a specific route and the experiences of his cousin, a Nantucket whaler, to decipher the Gulf Stream's flow. Many captains instantly grasped the value of the chart. Using simple thermometers to locate the Gulf Stream's warm waters, they took advantage of its swift flow on easterly voyages and avoided it on the return. Captain James Cook, who started

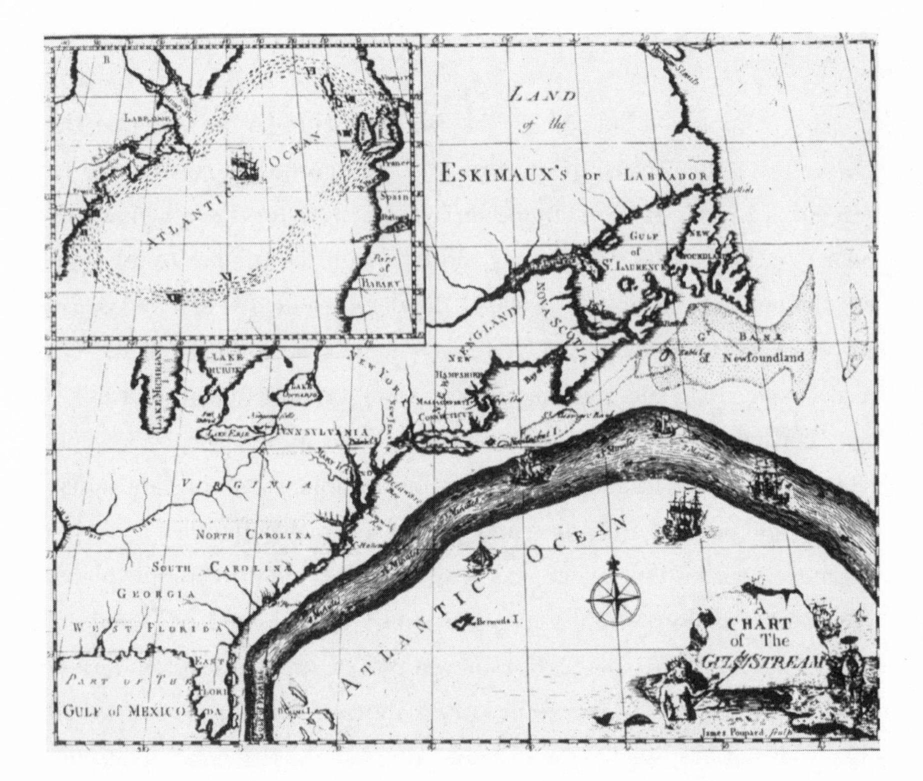

Figure 15 Benjamin Franklin's 1769 chart of the Gulf Stream. (Courtesy of NOAA.)

out as an officer in the British Royal Navy, commanded three major ocean voyages, during which he collected data on the sea's geography, geology, currents, tides, water temperature, and fauna. Cook's numerous and extraordinary achievements, included the confirmation of a great continent in the south, Antarctica, and the charting of much of the Pacific Ocean.

In the 1800s, scientific information about the sea was eagerly gathered as interest in trade, shipping, travel, fisheries, and national security rose. Governments, businesses, scientific societies, and wealthy individuals began to sponsor ocean research, and it became a time of great discovery—and in some cases, misconception. A crude device was built to measure the ocean's depths, little more than a very long rope with a weight attached. Sir John Ross and his nephew, Sir James Clark Ross, used this simple device to successfully, though tediously, measure ocean depths of some 2 to 4 kilometers. Dr. Alexander Marcet, a London physician at the time, noted that the proportion of chemical ingredients in seawater is constant throughout the world's oceans. Charles Darwin traveled aboard the HMS *Beagle* to South America and the Galapagos Islands, where he made observations that led to his theory of evolution. And Sir Edward Forbes sampled the ocean with towed nets and noted that the deeper and deeper he went, the fewer species he collected. He logically concluded that the deep sea must be devoid of life, or azoic. Forbes's azoic theory gained such widespread acceptance that even the few reported occurrences of life below 600 meters were ignored or discounted as human error.

In the mid-1800s, Matthew Fountain Maury, a lieutenant in the U.S. Navy, began a serious study of the winds and currents of the sea. He was convinced that he could identify patterns of wind and water that would allow for more efficient ocean travel. At the time, vessels that sailed from the United States to the Cape of Good Hope at the southern tip of South

America were known to traverse the rough waters of the Atlantic Ocean a tortuous three times. Maury sent out a request for sea captains and navigators everywhere to lend him their logbooks so that he could compile the wealth of information they contained. He later cajoled the ocean-going community, including those aboard naval ships, into making simple wind and ocean observations during voyages. Maury then produced a set of charts that helped to shorten ocean routes everywhere. For instance, the passage between New York and Rio that had always taken about 55 days could be completed in just 35 days. In 1855, Maury, often considered the father of oceanography, published the first textbook in oceanography, *The Physical Geography of the Sea*. The book, a compilation of the sea's currents, wind, temperature, depth, chemistry, and organisms, was an instant bestseller; it was translated into several foreign languages and was reprinted five times in its first year of publication.

During operations at this time to lay or repair the first transatlantic telegraph cable, samples of deep-sea mud were often collected or accidentally recovered. Sometimes the mud was preserved in alcohol and saved in small jars. In 1857, the preeminent British biologist, Thomas Henry Huxley, examined mud samples collected during a cable-laying cruise from Ireland to Newfoundland. In every sample he examined, Huxley found a grayish-white slime coating the mud. When stirred, the gooey substance separated into long strands of slime. At first, he thought that the gelatinous material was part of a giant deep-sea creature, but later he concluded that it must be a remnant of the primordial ooze, the organic slime from which life first arose. Huxley named the goo bathybius (from *bathy,* meaning "deep") and traveled throughout the world giving lectures and papers on its wondrous properties and significance. Bathybius would have but a brief moment of fame.

The latter half of the 19th century brought more revelations from the ocean's depths and the debunking of several theories. The azoic theory was the first to go. During repairs to a telegraph cable lying at 2000 meters within the Mediterranean Sea, a living deep-sea coral was recovered and quickly dispelled the azoic hypothesis.

In 1872, under the leadership of Charles Wyville Thomson and funded by the Royal Society of London, an ambitious and unprecedented voyage to study the oceans began. The *Challenger* Expedition was to be the first major cruise whose sole purpose was to study and explore everything about the sea—a daunting task even by today's standards. The guns of a warship were replaced with scientific gear and a crew of 240 was gathered. They packed 220 kilometers of line for measuring depth and some 20 kilometers of cable for collecting samples of sediment and marine life. The main deck was converted into a laboratory and workroom, and space was cleared for the storage of sample jars. Wine was brought aboard to drink and to use in sample preservation.

It took nearly three and a half years for the HMS *Challenger* to circumnavigate the globe and complete its mission. Each time the ship hove to and deployed sampling gear, it took an entire day to plumb the ocean's depths. Hours would be spent lowering a net or dredge (typically a square metal frame with a bag attached) to drag the sea for its treasures. Often the sampling device would break, get stuck, or come back empty (very similar to most ocean-sampling endeavors even today). Soon the sailors referred to the operation of dredging the seafloor as "drudging." Seasickness and bad weather plagued the expedition. But the scientists aboard continued to be thrilled by each and every sample wrung from the sea. Through the steadfast dedication of the crew and the passionate leadership of Thomson and the other scientists aboard, the voyage was completed and brought to light an extraordinary amount of new information about the sea. They measured depths reaching 936 meters (26,000 feet), mapped the distribution of seafloor sediments, and discovered 4017 new marine species. At the end of the voyage, the shelves were bursting with sample jars filled to the brim with marine creatures. From some of the exotic ports of call, they even brought back two large tortoises, spiders, a fur seal, and a young goat, not to mention the stowaway ants, cockroaches, moths, crickets, centipedes, and rats (Guberlet, 1964). A chemist aboard the ship also discovered that when alcohol (wine) used in sample preservation was mixed with seawater, a slimy white precipitate formed. The infamous bathybius was unmasked: it did not come from the seafloor, as Huxley thought, but was an accidental chemical by-product produced by the simple mixture of two common liquids.

After the *Challenger* expedition, oceanographic exploration was changed forever. Ships took to the sea to study the ocean, explore its depths, and reveal its creatures. In the United States, marine laboratories such as Woods Hole Oceanographic Institution and the Scripps Institution of Oceanography were established, along with government agencies concerned with the oceans: such as the National Marine Fisheries Service (NMFS) and the National Oceanic and Atmospheric Administration (NOAA). The sinking of the RMS *Titanic* and World Wars I and II sparked great advances in marine technology. The military needed to understand many aspects of the sea, including its thermal structure as it pertained to hiding and detecting submarines, the nature of waves and breakers at potential landing sites, and the behavior of sharks around stranded crew members.

Oceanographic studies later became more site- and topic-specific. Cruises tended to be short, expensive, and narrowly focused. Humans also began to enter the sea and observe its underwater wonders firsthand. Two things soon popularized ocean science: the wonderful adventures of Jacques Cousteau and the campaign to save the whales. Today, research cruises are even more expensive, so they are fewer in number and collaborative in nature. Typically, scientists from a diversity of marine disciplines work together and cover a broad range of oceanographic topics per cruise.

Our understanding of the oceans has progressed immensely over the last century, thanks to human curiosity, ingenuity, and a revolution in marine technology. Today, to study the oceans we have at our fingertips orbiting satellites, sophisticated computer models, highly sensitive and precise instruments, unmanned autonomous and remotely operated underwater vehicles, submersibles, undersea laboratories, and molecular research techniques (Color Plate 7). Satellite technology, submersibles, and remotely operated vehicles are allowing us to examine the ocean on a larger and deeper scale than ever before. Advances in molecular genetics and the study of small-scale phenomena are giving us a better view of the very smallest aspects of the sea. Yet, we have explored only a fraction of the ocean's depths. We are just beginning to understand the interactions between the ocean, the underlying Earth, and the climate and we still know precious little about the creatures that live in the sea's salty waters.

The Physical and Chemical Ocean

Over time the physical and chemical nature of the sea has evolved as a great balancing act that exhibits amazing consistency despite constant change.
—*Sylvia Earle*

ON TODAY'S EARTH, we tend to think of distinct oceans—the Pacific, the Atlantic, the Indian, and the Arctic. In actuality, there is only one global ocean connected through currents of air in the atmosphere and water in the sea. The composition of seawater seems to vary from place to place, but in reality the proportion of its constituents stays remarkably constant. It is the very special nature of this liquid we call water and its salts that literally move and breathe life into the sea.

Sea Salts and Water

The chemical nature of water and salts in the sea plays a crucial role in the ocean's flow and allows life to exist on Earth. Our own blood has a chemical composition similar to the seawater from which our ancient ancestors first emerged. Throughout human history, the sea's salts have also been harvested, used in trade, and valued as a food preservative or flavor enhancer.

Where does the sea get its salt, and why doesn't it change over time?

The saltiness of the sea generally refers to the amount of dissolved inorganic minerals (salts) in the ocean, in scientific lingo this is called salinity. Within seawater, dissolved salts are in the form of ions, or charged particles. The most common ions—the major constituents of seawater—are chloride (55 percent by weight), sodium (31 percent), sulfate (8 percent), magnesium (4 percent), calcium (1 percent), potassium (1 percent), and bicarbonate, bromide, boric acid, strontium, and fluoride (all less than 1 percent). The ocean also contains dissolved gases (carbon dioxide, nitrogen, and oxygen), nutrients (silica, nitrogen, and phosphorous), and minute or trace amounts of iodine, iron, manganese, lead, mercury, and gold.

There are three main sources of the sea's saltiness: weathering of rocks on land, volcanic gases, and circulation at deep-sea hydrothermal vents. When water combines with carbon dioxide it becomes acidic. Consequently, water vapor condensing to form rain in Earth's atmosphere, which contains carbon dioxide, tends to be slightly acidic. Rain that falls on the land, even if only slightly acidic, effectively dissolves rocks and sediments in a slow process we call weathering. Calcium carbonate rocks are particularly vulnerable to weathering by acidic rainwater. This is why the pyramids and Sphinx recently required a face-lift—because they are made of rocks built from the calcium carbonate shells of ancient foraminifera. Runoff and rivers carry the products of weathering, dissolved minerals, from the land to the sea. But if we compare the composition of river water to seawater, we will notice several distinct differences. River water has less chloride and calcium, and more magnesium than ocean water. Volcanic eruptions that spew gas rich in chlorine and sulfate from Earth's interior account for some of the missing constituents, but until recently, scientists were puzzled by the ocean's abundance of calcium and lack of magnesium. Neither river flow nor volcanic outgassing could explain the concentrations found in seawater. The mystery was solved with the discovery and study of circulation at deep-sea vents.

Deep-sea vents or chimneys occur along midocean ridges where plumes of mineral-rich superheated water erupt from fissures in the seafloor. Heated by molten material below, the temperature of the water emanating from an active hydrothermal vent field can range from a warm 25°C (77°F) to a fiery 400°C (752°F). The intense pressure at vent depths some 2500 meters below the sea surface allows the water temperature to rise above its boiling point and remain as a liquid—hence the term *superheated.* In some areas, superheated water escaping from the vents becomes trapped under ledges. Because the hot vent water is less dense than the surrounding cold seawater, buoyancy continues to drive it upward. In video footage, superheated water slowly escaping from beneath a deep-sea ledge appears as a shimmering upside-down waterfall.

Scientists examining the chemistry of seawater as it circulates through deep-sea vents and fractures made several startling discoveries. The rate of flow was much faster than ever expected and some now estimate that the entire volume of the oceans may circulate through the underlying oceanic crust in 10 million years or less (Humphris and McCollom, 1998; Baker and McNutt, 1998). In addition, chemical interaction with the underlying molten material causes circulating seawater to lose magnesium and gain calcium. At last, one of the sea's chemical mysteries was solved; here was the explanation for the lack of magnesium and the abundance of calcium in the sea. Circulation through deep-sea vents clearly plays a major role in the ocean's chemistry and heat flow.

On average, 1 kilogram of seawater contains 35 grams of salt, thus having a salinity of 35 parts per thousand (ppt). Throughout the oceans, the salinity of seawater can vary, but the proportion of its salts always stays the same. In other words, while salinity may vary between 30 to 37 ppt, it always has the same ratio of elements (55 percent chloride, 31 percent sodium, 8 percent sulphate, etc.). This was an important finding for researchers because it meant that the overall salt content could be determined if the amount of one of its constituents was known. Traditionally, chemists measured seawater's chloride content to determine salinity. Salinity can also be determined by measuring the conductivity of seawater (the more salt, the more conductive it is). Today, scientists use instruments that compare the electrical conductivity of a water sample to that of a standard seawater sample. Sensors to measure salinity, temperature, and pressure (for depth) are often combined in an instrument package called a CTD (conductivity, temperature, and depth). The sensors, along with a computer to record and store the data, are commonly placed inside a short, cylindrical, watertight housing and lowered on a cable from a ship to measure the properties of seawater with depth (depth is measured by pressure change).

Water is so common on Earth's surface that we often take its special properties for granted, but without it, neither the planet as we know it nor humans would exist. Water can exist on Earth as a liquid, a solid, and a gas. How many other substances can do the same? Simple though it is, the water molecule—two hydrogen atoms and one oxygen atom—has some truly amazing properties. It has a large, positively charged oxygen atom on one end, and two negatively charged hydrogen atoms on the other. These oppositely charged ends act like a magnet, the positive side attracting particles with a negative charge and the negative side attracting particles with a positive charge. This is a powerful force in the wet world; water dissolves more substances in greater quantities than any other liquid. When water molecules move about freely, they are water vapor—a gas. An increase in temperature will cause the water molecules in the gas to move around faster, causing it to expand and become less dense. In cooler temperatures, the molecules slow down and some form weak bonds between their hydrogen atoms, thus forming a liquid—water. The ocean can store great amounts of heat, because lots of energy must be added before the water molecules break their bonds and evaporate as water vapor.

In really cold temperatures, all of the hydrogen atoms within the water molecules attach to each other in a six-sided ring and form a solid—ice. Because the angle between oxygen and hydrogen in the ice crystal is greater than in its liquid form, it is slightly more open and therefore less dense; this is why ice floats in water. When seawater freezes, salt crystals cannot quite squeeze into the ice structure, so they get excluded and the salinity of the surrounding water increases. The attraction of hydrogen atoms in water also produces a high surface tension. The only liquid with a higher surface tension is mercury, well-illustrated by its ability as a liquid to form small beads and roll around.

Some marine organisms, such as the only open-ocean insect, Halobates, or water strider, use water's surface tension to live on and move over the very skin of the sea. The water strider has tiny water-repellent hairs on its legs and feet that help it to tiptoe across the water's surface and very long legs relative to its body size to spread its weight out over a large area. For humans to do the same, our legs and arms would have to be about 8 kilometers long!

Probably one of the most important properties of seawater is its density. The density of seawater increases when either temperature is lowered or salt is added; conversely its density decreases when heated or fresh water is added. An increase in salinity will raise seawater's boiling point and lower its freezing point and vice versa.

Throughout the ocean, the composition of seawater stays relatively constant due to the eternal inflow and outflow of its elements. Just as salts are put into the sea by weathering, volcanic eruptions, and circulation in deep-sea fractures, they are removed by processes such as deposition on or within the seafloor, evaporation, and sea spray. Water itself is also continually coming and going within the sea. Each year, evaporation removes huge quantities of water from the ocean's surface. Rain, snow, and river flow return the water and bring salts back to the sea. Climate change can alter the amount of water cycled in the air-land-sea system and cause global sea level to rise or fall. Water is by far the most important substance on our planet—quite literally the fountain of life.

The Ocean in Motion

Like a huge, twisted conveyor belt the oceans transport warm water on the surface and cold water below. It is a system of constant give and take, driven by the influences of climate, gravity, pressure, and ultimately, by the sun.

The Sun's Rays

Earth's winds are an old and fundamental component of our planet; they transport heat, create waves, and drive currents on the ocean's surface. Winds are created by the powerful, but uneven, heating of the planet by the sun. It is a process that begins when the sun's incoming rays enter Earth's atmosphere. Here, the sun's rays (energy) are absorbed, reflected, or scattered by ozone, clouds, dust, and various gases. The solar radiation that passes through the atmosphere strikes Earth's surface and is again, absorbed, scattered or reflected. Where lots of the sun's energy is absorbed, Earth's surface warms; where energy is less intense or reflected, the surface remains cool. The resulting differences in temperature on the planet's surface create rising and falling plumes of air in the atmosphere that generate our earthly winds.

What happens to the sun's incoming energy depends greatly on its wavelength. Solar energy appears as light, feels like heat, and travels as waves. Within the atmosphere and ocean, long wavelengths (the distance between peaks of energy) of light tend to be absorbed by particles, while short wavelengths are scattered. Blue light has a short wavelength and on most days we can easily see evidence of its scattering—the color of a clear sky. Both blue and green light are relatively short-waved, while red and yellow are long-waved. Glowing sunsets of red and gold occur when dust or other larger particles in the atmosphere scatter the longer wavelengths of light. A similar process has lent the sea the apt name, Big Blue. As light hits the sea surface, water molecules and other materials in seawater tend to absorb longer wavelengths and scatter shorter ones. Consequently, blue and green light are scattered and penetrate deep in the sea, while red and yellow are absorbed, usually within the top 20 meters (65 feet) or so. If someone is cut while SCUBA diving at depths below about 20 meters, the liquid oozing from the wound appears—in horror-movie splendor—blue-green in color. (Personal experience with this will instill the concept of red-light absorption for life.) Under the glare of an underwater light the blood returns to its normal red color. But when the light is turned off, the blood again appears blue-green; all red light has been absorbed so we cannot see red at depth. In coastal regions, the sea often appears green or brown, because particles of sediment and organic matter reflect and scatter green and yellow wavelengths of light.

At Earth's surface, another factor that determines whether the sun's energy is absorbed or reflected is the surface's reflectivity. At high latitudes, snow and ice create a light-colored surface that effectively reflects the sun's rays and keeps the region cool, like a light-colored shirt in the tropical sun. In contrast, where Earth's surface is relatively dark, such as the ocean, solar radiation is mostly absorbed and creates warmth. Nearly

65 percent of the visible light hitting the sea surface is actually absorbed within the top 1 meter, thus elevating surface-water temperatures. The sea is a natural heat storage facility; it not only absorbs much of the incoming radiation but also tends to retain its warmth. Compared to the ocean, land not covered by ice or snow has a tendency to heat and cool relatively quickly.

In coastal regions, windsurfers and sailors know well the daily oscillations of the wind that create a regularly occurring sea breeze. During the day, the sun rapidly heats the land relative to the nearby sea. Air warmed over the heated land becomes less dense and rises. The rising air is replaced by cooler air sucked in from the adjacent ocean and by midday or late afternoon an onshore breeze develops (Figure 16a). At night, the land cools off more rapidly than the sea. Air cools and descends over the land, while it rises over the relatively warm coastal ocean. The flow of the wind reverses, and at night a breeze blows offshore (Figure 16b). In the tropics a line of white, puffy clouds typically forms just offshore during the night as air and water vapor rise over the relatively warm ocean.

Most of the sun's radiation that passes through Earth's atmosphere and hits the surface is short-wave; the long-wave radiation is absorbed by atmospheric clouds, dust, and gases. As Earth is heated it also emits radiation—so some of the sun's energy that hits the planet is re-emitted back into the atmosphere. However, the outgoing radiation is longwave. When

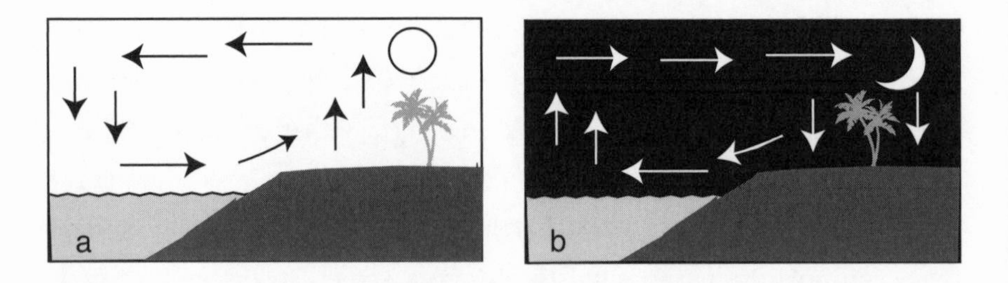

*Figure 16 Sea breeze circulation (*a*) in the daytime and (*b*) at night.*

this re-emitted, long-wave radiation enters the atmosphere, it is there absorbed by clouds, dust, and gases. The absorption of the outgoing radiation heats the atmosphere further; this is known as the *greenhouse effect*. Because carbon dioxide is one of the main absorbers in the atmosphere, activities on Earth, such as fossil-fuel burning and deforestation, which release carbon dioxide into the atmosphere, can cause the climate to warm.

Finally, the angle at which the sun's radiation hits Earth's surface also affects the amount of incoming energy received. Because Earth is round, the amount of solar radiation that hits the surface varies across its curvature. At the equator, the sun's energy strikes straight on or perpendicularly, whereas at the poles and at higher latitudes, it hits at a steep angle and is spread out over a larger area. Consequently, heating is much greater at the equator than at the poles. Why then, given the latitudinal difference in heating and the high reflectivity at the poles, are not the tropics overheating and the poles plunging into a permanent deep freeze?

Earthly Winds

On an imaginary, geographically and politically simple Earth—no land, no people, no spinning about its axis—heating from the sun would create rising air at the equator and sinking air at the poles. When a substance warms, becomes less dense, and rises within its surroundings, it is called *convection*. Convection is a very efficient means of heat transfer and occurs in the planet's atmosphere, oceans, and interior. To envision the process of convection, think of the colorful motion in a Lava Lamp. At the lamp's base, heat from a light warms colored wax in a surrounding oil mixture. As the wax heats up it becomes less dense than the oil and rises. Near the top of the lamp, away from the source of heat, the wax cools and sinks back to the bottom of the lamp; this is convection.

On our simple Earth, air rising near the equator sucks air in and up from the surface, and sinking air at the poles spreads down and out over the surface. This generates a kind of circular wind pattern, with air near Earth's surface blowing from the poles (high pressure) to the tropics (low pressure) and in the upper atmosphere, from the tropics to the poles. Unfortunately, or maybe fortunately for us, Earth is not quite so simple. The addition of the planet's rotation and land masses creates a multicellular system of atmospheric circulation, with at least six circular cells spanning Earth's diameter (Figure 17). And unlike on our simplified planet, surface winds on Earth blow not in a north-south direction, but in an east-west direction. Why?

The Coriolis Curve Ball

To explain this, let us set up a completely hypothetical situation. A raging war on drugs has impelled the U.S. military to fire a missile from a secret

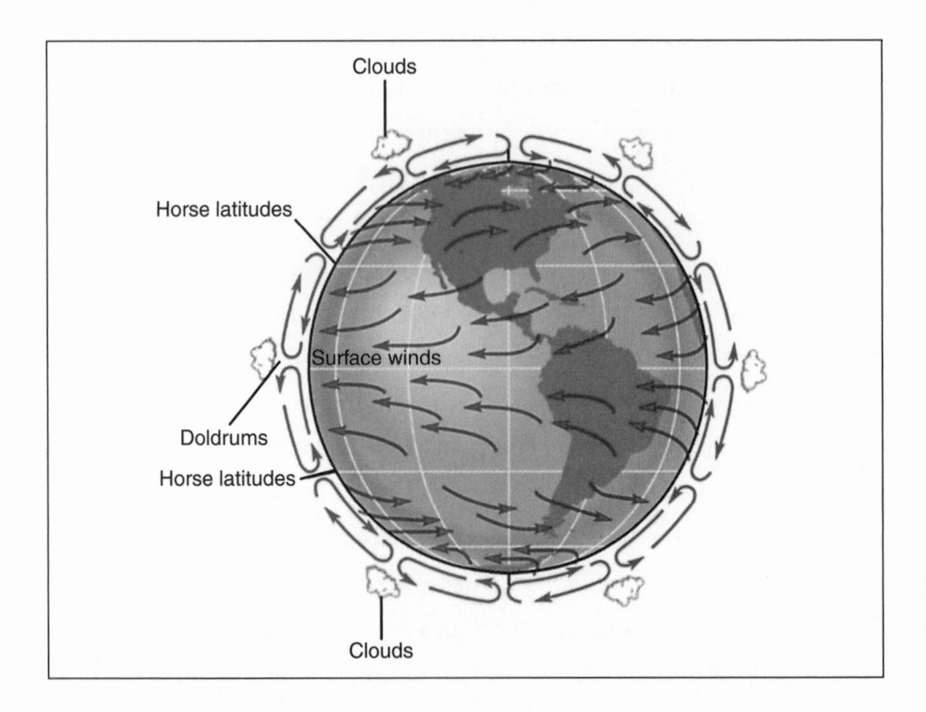

Figure 17 Atmospheric circulation and direction of surface winds over the globe. (Reprinted by permission of John Wiley & Sons, Inc. from Exploring Ocean Science, *Keith Stowe, © 1996, John Wiley & Sons, Inc.)*

location outside of Washington, D.C., into a drug lord's compound in the jungles of Columbia, near the equator. Under the cloak of night and the highest of secrecy, the missile is launched. However, it lands not in Columbia, but on one of the small islands in the Galapagos Archipelago, some 500 miles west of South America (Figure 18). Numerous endangered tortoises and much of their fragile habitat are annihilated; environmental groups vigorously protest at the White House and Ecuadorian officials are enraged. The United States takes quick action to fire the missile technician. Much to his chagrin and embarrassment, the technician realizes his mistake; he forgot to program the missile guidance system to adjust for the Coriolis effect.

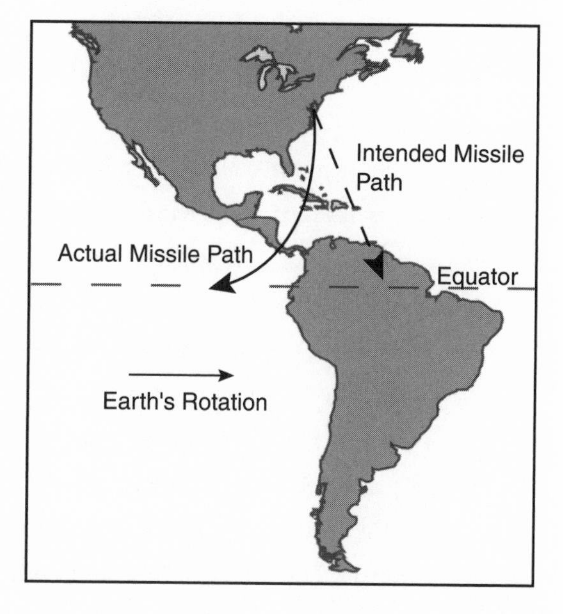

Figure 18 Apparent deflection of a missile due to the Coriolis effect.

Coriolis is an apparent force that acts on bodies that move over Earth's surface but are not frictionally coupled to it. It causes a deflection to the right in the Northern Hemisphere and to the left in the Southern Hemisphere. Before we explain Coriolis, let's take a step backward and consider Earth's rotation. In a single day, the entire planet spins once around to the east. Because the circumference of Earth is greater at the equator than at the poles, the Earth must move faster at the equator than at the poles for both places to rotate completely in the same amount of time. Earth moves approximately 1600 kilometers per hour (1000 mph) at about 2 degrees of latitude and 800 kilometers per hour (500 mph) at

60 degrees of latitude. If an object or mass is frictionally coupled to Earth's surface, it moves along with the planet's motion. To understand this, think of your hand sliding a piece of paper across a desk; the paper moves along with your hand because it is frictionally coupled—essentially, it is stuck to your hand. If your hand is slightly above the piece of paper and you move it in the same way, the paper does not move because it is no longer frictionally coupled to your hand. Frictional coupling between Earth's surface and moving air or water is relatively weak. So, what happened to the missile without Coriolis accounted for?

When the missile was sitting on the surface it was frictionally coupled to the planet. As soon as the missile became airborne, its frictional coupling weakened. As the missile left Earth's surface from somewhere near Washington, D.C., it was imparted with an eastward speed equal to that of Earth's rotation at that latitude. But because the target was near the equator, it was spinning to the east faster than the missile launch site. By the time the missile got to the latitude of Columbia, the target had already rotated past, and the missile struck behind or west of its intended target. The technician should have added an acceleration eastward to account for the faster rotation at the equator. Drawing an arrow starting at the launch site and following the missile's path to the landing site, it would appear to curve toward the right (Figure 18). This apparent deflection is called the Coriolis effect.

The Coriolis effect also tends to increase toward the poles and has a similar effect, though slightly more difficult to envision, on objects moving to the east or west. If an object is moving to the east—the same direction as Earth's rotation—conceptually the object gains speed and slips toward a lower latitude with a faster rotation to the right in the Northern Hemisphere and to the left in the Southern Hemisphere. If the object is moving west, against Earth's spin, it slows and moves poleward, again to the right in the Northern Hemisphere and to the left in the Southern

Hemisphere. The Coriolis only affects relatively large-scale phenomena that are not bound by friction to the planet. If you're driving down the highway, you won't see cars constantly swerving to the right while their drivers battle the Coriolis force.

Now back to our question about the winds at Earth's surface. By now, the answer should be clear: global wind patterns blow in an east-west direction because of the Coriolis curve ball. In a rather simplistic summary, the differential heating of Earth's surface results in cells of rising, sinking, and horizontally flowing air that are affected by the Coriolis force, causing a large-scale system of winds that blow mainly east-west over Earth's surface. At the equator and at about 30 degrees north latitude and 30 degrees south latitude, surface winds are notoriously weak. In these areas, air movement is more vertical than horizontal (Figure 18). Sailors refer to the low-wind equatorial region as the doldrums, from the hours of boredom endured while waiting for a good sailing breeze to develop. The area at 30 degrees north and 30 degrees south latitudes are nicknamed the horse latitudes, supposedly because in the absence of wind, sailors used to dump horses into the sea to lighten their load.

Wind over Water

Picture the ocean as made up of a series of stacked layers of water, each loosely connected to the other by friction. Wind over the ocean's surface drags the surface water and each successive layer below it. However, as depth increases and the water layers get farther and farther away from the force of the wind, each layer is progressively moved less and less so that at some depth within the sea there is little, if any, motion created directly by the wind. This usually occurs at a depth of about 100 to 200 meters. The region from the surface to the depth where wind has little direct influence on the ocean is called the *mixed layer.* Again, if we were on a simple, nonspinning Earth, the ocean's surface currents would move in the same direction as the wind. But then again, the Coriolis curve ball is thrown into the equation.

In the 1890s an oceanographer named Fridtjof Nansen led an expedition across the Arctic ice in a specially designed vessel, the *Fram*. The ship was actually frozen into the ice and allowed to drift. After a very long, cold year and some chilly adventures on the ice, Nansen observed that the ice movements were not parallel to the wind, as expected, but at an angle some 20 to 40 degrees to the right of the wind. Later, a graduate student, V. W. Ekman, was asked to come up with a theory to explain why ice and ocean water apparently flow to the right of the wind (in the Northern Hemisphere). Amazingly enough, overnight Ekman apparently came up with the following explanation: In our layered ocean, as each section of water begins to be dragged by the force of the wind, it is also acted on by the Coriolis force. In the Northern Hemisphere, each layer is successively deflected to the right, spiraling downward in what is now called the Ekman spiral (Figure 19). If the direction of flow is averaged over the entire mixed layer, the net transport is about 90 degrees to the right (left in the Southern Hemisphere) of the wind and is called Ekman transport. Ekman transport is very important along continental margins, where it can cause coastal upwelling.

Coastal Upwelling

In several regions of the world wind blows parallel to the coast and Ekman transport causes the surface water to flow offshore. To replace the offshore-flowing surface water, cold, nutrient-rich water wells upward from below; this is called coastal upwelling (Figure 20). Areas of coastal upwelling are some of the most fertile regions in the sea. Here, phytoplankton (floating plants) use upwelled nutrients to photosynthesize and grow in prolific numbers. Then, as long as upwelling continues, zooplankton (floating animals) and small fish come to dine and prosper

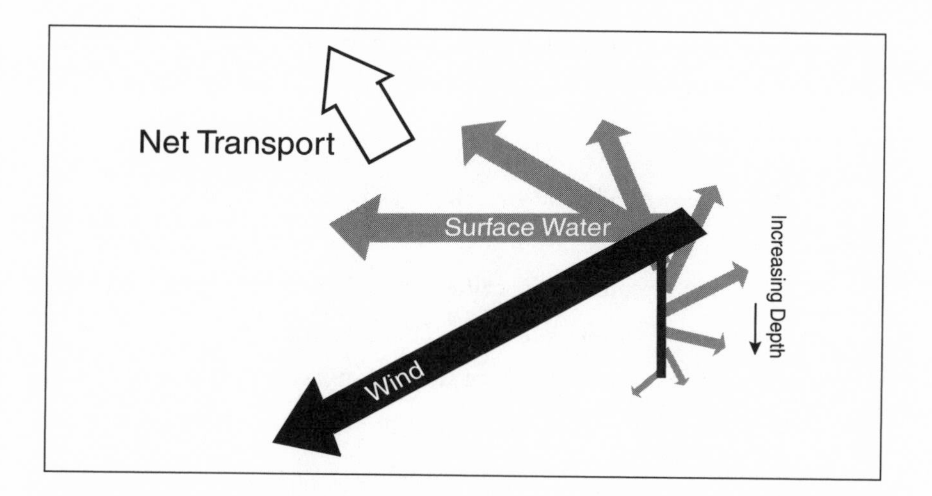

Figure 19 The Ekman spiral and net transport of water to the right of the wind in the Northern Hemisphere.

on the constantly replenishing smorgasbord of food. Off Peru, along South America's west coast, northerly blowing winds cause upwelling and create one of the richest anchovy fisheries in the world. Coastal upwelling also occurs off the coast of California and, during the summer, off the northeast coast of Africa. During years when El Niño is particularly powerful, coastal upwelling shuts down and major fisheries typically collapse.

Upwelling also occurs within the equatorial region of the sea and in the southernmost ocean (north of Antarctica). Near the equator the trade winds blow from east to west and Ekman transport causes the surface waters to deflect to the north and south, away from the equator.

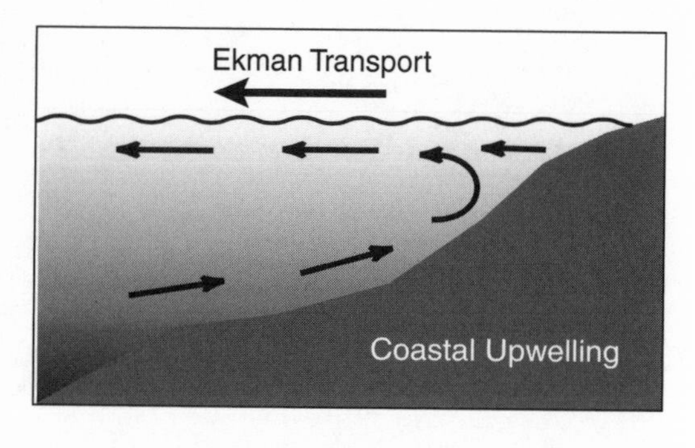

Figure 20 Coastal upwelling.

Cold, nutrient-rich waters well upward from below and create a narrow equatorial zone of fertile water rich with life.

Ocean Gyrations

The combined effects of wind-driven motion at the sea surface and the distribution of land cause the surface waters of the global ocean to move in a series of large circular flows, called gyres (Figure 21). These gyres are distinct features within the world's oceans; they are separated by flow at the equator and play a major role in the transport of heat in the sea and air. Circulation in the North Atlantic Ocean illustrates how gyre systems form and operate throughout the sea today.

Winds over the northern half of the North Atlantic tend to blow toward the east, and in the south, the trade winds blow toward the west. In a debate that is sure to confuse all students of oceanography, oceanog-

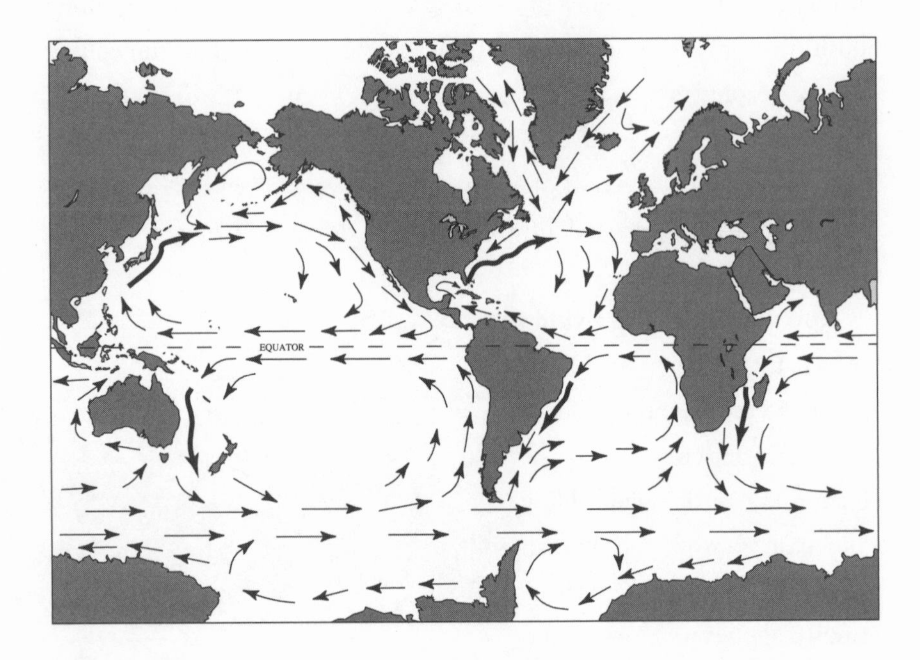

Figure 21 Major surface currents and gyre circulation in the world's oceans.

raphers name winds and currents based on where they are flowing *to,* and meteorologists name them on where they are coming *from.* So the trade winds that blow from the east are easterlies to the meteorologist and westerlies to the oceanographer. With the winds blowing *from* the west in the north and east in the south of the North Atlantic, Coriolis and Ekman transport cause the surface water to flow toward the middle of the North Atlantic Ocean. The convergence of surface water causes a literal pileup of water in the middle, in an area known as the Sargasso Sea.

In the ocean the surface forms small rolling hills and valleys that help to drive currents. Using satellite altimetry, we can now precisely measure relatively small changes in sea surface height from space (Color Plate 8a). Measurements of sea surface height show that within the central Sargasso Sea there is about a 1-meter-high pileup of water. Floating materials such as plastic, tar, and sargassum, a floating seaweed, tend to accumulate in the water converging in the middle of the Sargasso Sea. According to historical accounts, sargassum once formed dense mats across the central region of the North Atlantic, hence the name Sargasso Sea.

Sargassum can live either free-floating in the open ocean or attached to the seafloor in shallow, warm-water regions. Small, berry-shaped bladders speckle the plant and keep it afloat. With little protection and few objects to cling to in the open ocean, many small organisms live within the small clumps and great rafts of sargassum in the Sargasso Sea. The sargassum fish with its mottled-brown coloring and seaweed-like growths looks so much like the seaweed itself that it is often hard to distinguish fish from plant. Though small, the sargassum fish is a fierce and voracious competitor. If two are placed in a fish tank, soon there will be only one. The victor may bloat guiltily to double its normal size as a result of consuming its fellow fish. Also common in the Sargasso Sea are the amazing flying fish. These fish propel themselves out of the water and glide effortlessly over the surface using their tails as rudders and outstretched fins as wings. Flying fish have been known to fly onto boat decks, through open portholes, into air intake tubes—even into the face of a hapless sleeper.

Surface water continually piles up at the center of the Sargasso Sea. Consequently, a pressure gradient forces water to flow outward beneath the surface pile. As water flows outward below, Coriolis comes into play and the moving water curves to the right. This process—surface water piling up, flowing outward, and to the right below the mixed layer—creates a large gyre of currents circulating clockwise in the North Atlantic (Figure 21). A similar pattern occurs in the South Atlantic except that because Coriolis acts to the left, the gyre circulates counter-clockwise. Ocean gyres also occur in the Pacific and Indian oceans, although the Indian Ocean system is modified by seasonal changes in the monsoon winds. Around the Antarctic, where no land boundaries exist to block flow, a globe-encircling, or circumpolar, current flows around the entire Southern Hemisphere. Additionally, beneath the westward-flowing equatorial currents lies an undercurrent going in the opposite direction. Typical open-ocean currents, not including boundary currents such as the Gulf Stream and its pacific counterpart, the Kurishio, flow at speeds of about 8 kph (5 mph).

Deep-Sea Circulation

Just as wind drives circulation at the ocean's surface, gravity drives flow in the deep sea. On average, the ocean is some 4 kilometers (2.5 miles) deep. Therefore, most of the ocean lies below the mixed layer. While we know quite a bit about circulation in the shallow sea, the slow flow of the deep ocean remains more of a mystery. Near the surface and in the mixed layer, oceanographers use drifters and stationary current meters to measure flow. Just below that, they use sea surface height, gravity, and pressure differences to calculate flow. But beneath these two areas lies the deep sea. Here, measuring or calculating flow is much more difficult. In fact, we know the least about the area, that makes up 90 percent of the total volume of the ocean.

Water motion in the deep sea is slow, driven by gravity and caused primarily by changes in the density of seawater. The colder and more salty the sea gets, the heavier and denser it becomes. For the most part, it is at the surface, the interface between air and sea, that temperature or salinity change. The cooling of the sea takes place when a chill wind blows over the surface or a cold air mass sucks the warmth out of the sea. An increase in salinity can occur with evaporation or the formation of sea ice. If the density increase due to these processes is sufficient, ocean water will slowly sink and flow downward until it reaches a level of equal density or the seafloor.

Almost all of the ocean's deep water forms through the effects of cooling and freezing at high latitudes. By far, the area that generates the most bottom water lies just south of Greenland in the North Atlantic. Here, the warm, salty waters of the Gulf Stream merge with cold waters flowing south around Greenland. When these waters collide they produce prodigious amounts of cold, salty water that cascades downward and spreads throughout the deep Atlantic. This deep-water mass is known as North Atlantic Deep Water and is produced in such great quantities that it essentially fills most of the Atlantic Ocean. Scientists have traced its flow past the equator and deep into the Southern Hemisphere. Near the Antarctic, North Atlantic Deep Water mixes with water flowing around Antarctica and then moves into the Pacific and Indian oceans. Little bottom-water actually forms in the Pacific or Indian oceans; most of it comes from the Atlantic. The very densest seawater forms during the southern winter beneath the Antarctic ice shelf. Here, the water is extremely cold and very salty, so it sinks all the way to the seafloor, spreads out, and flows northward, beneath the somewhat less dense, southerly-flowing North Atlantic Deep Water. However, Antarctic Bottom Water generally stays in the Atlantic Ocean because ridges on the seafloor block its path. Cold bottom water also forms during the winter in the Arctic, but because of the surrounding continents and seafloor ridges it remains within the Arctic Ocean basin.

Since there are few means of mixing water in the deep ocean, water masses tend to move as distinct layers flowing within the sea. Each water mass has a suite of characteristic properties, such as temperature, salinity, oxygen, and silica content. By identifying and tracking these properties with depth, oceanographers can trace water masses as they move throughout the ocean. One of the most common ways to sample water in the deep sea is to use a specially designed collecting device called a Niskin Bottle.

A Niskin Bottle is an ingenious and inexpensive piece of equipment (though getting it into the deep sea is very costly). The bottle, usually made out of thick gray PVC tubing, is attached to a cable, directly or as part of a larger sampling unit, and lowered into the sea. On the way down, both the top and bottom of the bottle are kept open. Once the depth to be sampled is reached, a triggering mechanism releases the lids and the bottle is snapped shut. In larger sampling units, triggering is done by computer, but if just one or several Niskin Bottles are used, then a weight, called a messenger, is clipped onto the cable and slid down to trip the closing mechanisms. With the seawater tightly sealed inside, the Niskin Bottles are then raised to the surface. Care must be taken when using Niskin Bottles to ensure correct setup and triggering. Also, because the lids snap shut with great force, there is much danger to the fingers if a bottle is accidentally triggered during handling. Using a CTD and numerous Niskin Bottles that trigger at various depths, researchers can sample at a location and with subsequent chemical analyses identify the different water masses present throughout the water column.

In between the surface and deep waters of the sea lies the intermediate ocean. In some places, water masses form and flow into the intermediate ocean, wedged between the warm waters of the surface and cold waters of the deep sea. In the Mediterranean Sea intense evaporation creates a very salty, warm intermediate water mass that flows out through the Straits of Gibraltar, beneath less salty, incoming surface water. Even though it is warm, the Mediterranean water is so salty that when it enters the North Atlantic it spills downward to a depth of about 1000 meters, where colder water is of an equal density. Sandwiched between the upper and lower layers of the ocean, Mediterranean intermediate water forms a salty liquid avalanche spreading down and out. Relatively recent research suggests that, much like the Gulf Stream, Mediterranean intermediate water flowing within the North Atlantic meanders and creates circular swirls of flow, called eddies. Scientists have nicknamed these Mediterranean water eddies as Meddies, and have tracked them for up to 7 years as they slowly drift within the intermediate depths of the sea (Richardson, 1993).

To measure the ocean's flow researchers and government agencies have released hundreds of surface-drifting buoys and established moored stations throughout the world. The drifters travel along with currents and are tracked via orbiting satellites. Every few days a transmitter on the drifter sends data to a satellite, which then sends a signal to a receiving station on land. There, the data are converted into information such as latitude, longitude, or sea surface temperature. A scientist can obtain this information, often through the Internet, and plot or track where the drifter has traveled and any of the properties it has measured. Moored stations work in a similar way, except that they measure water properties, such as current speed or temperature, in one location over time. Soon scientists will be able to measure both currents and waves at the sea surface from space. NASA's recently launched QuikSCAT satellite houses an instrument known as the Sea Winds scatterometer, which will use microwave radar to measure and track winds, surface waves, and currents. New data are expected to improve our understanding of ocean processes as well as ocean-related climate phenomena such as hurricanes and El Niño.

To very precisely measure the sea's flow in the horizontal as well as vertical direction, oceanographers today can also use sound waves via shipboard and stationary instruments called acoustic doppler current meters. And to track flow at depth, researchers can use specially designed drifters. The most recent version of this type of float or drifter is called Argo and is being used to trace ocean currents and make measurements at depths of up to 2000 meters. Upon release, Argo floats sink to a target depth equal to their own density and drift for up to 10 days. Then, programmed by an internal computer, they rise to the surface, recording temperature and salinity along the way. Once on the surface the Argo float radios its position and data to an orbiting satellite before returning to depth and another interval of drifting. Scientists expect that Argo floats can continue measuring and sending signals like this for up to 4 or 5 years. Together with surface drifters, satellite images, and all other available data, scientists hope to use the resulting data to create ocean "weather maps" to improve our understanding of the ocean and its interaction with the atmosphere, and to help create more accurate computer models of climate change.

An Ocean Conveyor Belt

The global circulation of the ocean is much more complicated than depicted here. Flow is undoubtedly interrupted, enhanced, slowed, and diverted on a regular basis. In fact, one of the greatest revelations of the last century may be that flow within the ocean is much more variable than ever believed, in both space and time. And yet, the big picture remains a sort of ocean conveyor belt or watery highway of heat transport (Figure 22). Warm surface waters driven by the forces of wind, pressure, and Coriolis tend to move in large circular gyres. Within these gyrating flows, heat is transported from the equator to the poles. At high latitudes, the chill of the air and water, the freezing of ice, and the evaporative winds cool and raise the salinity of the warm, newly arrived surface waters. Density increases, and surface waters plunge downward, spreading toward lower latitudes, and filling the sea with dense, frigid water. Water flowing along the bottom is often blocked or bumped off course by ridges, rises, and other underwater obstacles. But like a tourist seeking a winter reprieve, the deep waters move toward the tropics and begin to heat up. Some of the water warms and rises toward the surface, reentering the gyre system; here it once again heads toward the frigid north. Meanwhile, a portion of the deep water remains near the bottom

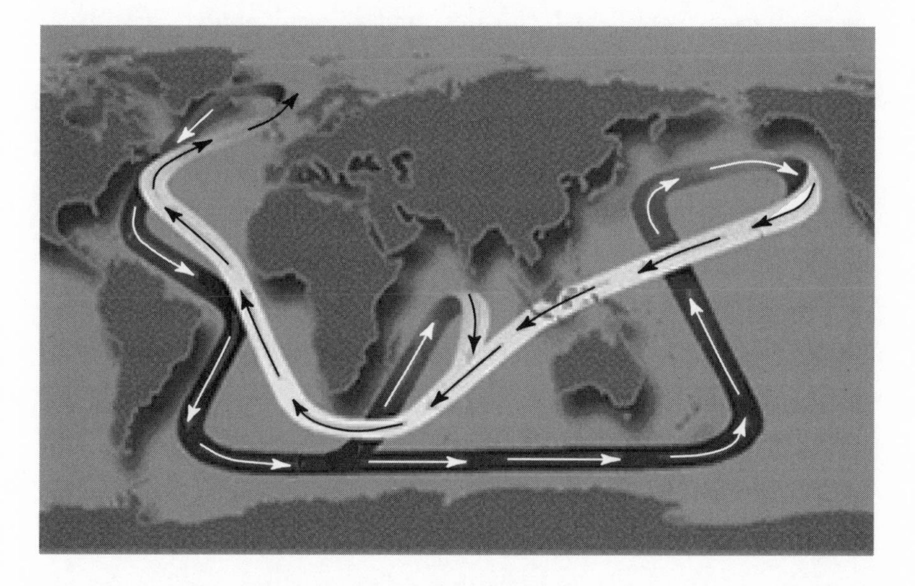

Figure 22 Simplified ocean circulation depicted as a giant conveyor belt. (Courtesy of Jet Propulsion Laboratory, NASA.)

and continues on its cold journey south, mixing with other deep waters and moving into other ocean basins. The ocean conveyor belt provides the planet with a second means of naturally regulating its heat distribution (the first being wind). The currents of wind and water circulating about the planet offset the sun's uneven heating of Earth and, in doing so, prevent the tropics from continually overheating, and the poles from plunging into a permanent deep freeze.

Regional Seas

Within the global ocean are smaller-scale features that play a supporting role in the big picture and greatly influence coastal environments. Partially enclosed, relatively large embayments of the ocean are called regional seas, or gulfs. Examples include the Gulf of Mexico, the Gulf of Maine, and the Caribbean Sea. Because they lie adjacent to land and have somewhat restricted flow, regional seas are often strongly influenced by river drainage, continental climates, and human inputs. Circulation within a gulf or sea may be controlled by local changes in depth, river inflow, wind, and ocean currents. Seasonal increases in river discharge can often be traced as a spreading plume of freshwater. For instance, using satellite imagery or water mass properties, the outflow or plume of South America's Orinoco River can sometimes be traced for hundreds of kilometers as it flows into the Caribbean Sea (Color Plate 9b).

The Gulf Stream

The narrow, swift flow and incessant wanderings of the warm Gulf Stream are one of the most dramatic and easily observed physical phenomena in the sea (Color Plate 10). For those who have personally experienced its changing nature, it can be a pleasant and fascinating flow or a hair-raising, stomach-heaving mass of choppy waves and racing water. The first chart of the Gulf Stream (Figure 15) made by Benjamin Franklin and his whaler cousin, Timothy Folger, shows a relatively large, river of water moving northward along the coast from Florida up to North Carolina, where it veers to the east, widens and continues across the North Atlantic. Considering the oceanographic tools of the time—little more than a thermometer and ship records—this is an amazingly accurate representation of the Gulf Stream's flow. Today, we have a much better picture of the Gulf Stream, why it occurs, how it flows, and its impact on the sea and climate.

The Gulf Stream is approximately 50 to 75 kilometers wide, about 2 to 3 kilometers deep, and flows at a rate of 3 to 10 kilometers per hour. It has been estimated that in some areas, the Gulf Stream transports somewhere on the order of 70 million cubic meters of water each second, about a thousand times the amount of water moved by the Mississippi River (Hogg, 1992).

One of the earliest explanations for the Gulf Stream was that the trade winds, constantly blowing from the east to the west, pile water up along the South American coast near the equator and cause a "downhill" flow of water to the north and south. While this is true, it does not sufficiently explain why the flow of the Gulf Stream and the other similar western boundary currents, like the Kurishio in the Pacific, are so swift and narrow. Oceanographers have now identified two other factors that intensify flow along an ocean basin's western boundary. The great ocean gyres, including that in the North Atlantic, are slightly offset to the west, so that the pile of water at its center is slightly steeper along its western side. Just as on land, the steeper the slope, the faster objects (in this case, water) fall downhill. In the North Atlantic gyre, think of water as flowing out from the central pile and moving to the right (Coriolis). Where the slope of the pile is steep, water moves faster than where the slope is gradual. On the western side of the gyres, where the sea surface slope is relatively steep, currents tend to be fast.

The other factor that intensifies the Gulf Stream is a bit more complicated, but a relatively easy way to think of it has to do with changes in the effects of Coriolis. The Coriolis force increases with latitude, so a water parcel moving poleward is increasingly affected by Coriolis the farther north it gets. This imparts a clockwise spin on the water parcel. Therefore, on the western side of the North Atlantic, both the velocity due to spin and the water motion itself are in the same direction, and the flow is intensified (Figure 23a). Along the eastern margin of an ocean basin, gyre currents flow toward the equator. However, the effects of Coriolis decrease toward lower latitudes, thus imparting a counterclockwise rotation on the water parcel. On the eastern side of the basin, then, the direction of the flow and spin are opposite or opposing, and currents tend to be relatively slow (Figure 23b). The combination of these three factors—wind pileup along the shore, the offset of the gyre with its steep western slope, and the increasing effects of Coriolis with latitude—cause the intensification of currents, like the Gulf Stream, on the western side of ocean basins. In general, western-boundary currents tend to be fast, intense, deep, and narrow, whereas eastern-boundary currents are typically slow, wide, shallow, and diffuse.

Much like an old river, the Gulf Stream meanders north and east, sometimes forming great curves, bends, loops, and rings. In rivers,

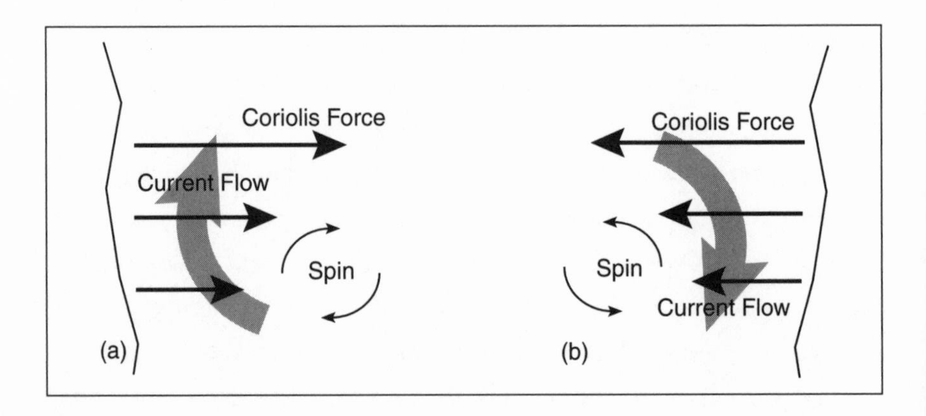

Figure 23 Clockwise spin and current flow (a) *in a western ocean boundary current and* (b) *opposing forces in an eastern ocean boundary current.*

where a sharp bend forms and gets cut off from the main flow an oxbow lake may be created. Similarly, a sharp bend or loop in the Gulf Stream can become separated from the main current and form rings. Gulf Stream rings have a warm- or cold-water core that is enclosed by a circular flowing remnant of the main current. When the Stream bends to the north, breaks off, and forms a ring, it is called a warm-core ring. They are usually 100 to 200 kilometers across and have a central core of warm subtropical water from the Sargasso Sea and cooler, outside waters that rotate clockwise. A meander that bends southward, breaks off, and traps cold water at its center forms a cold-core ring rotating counterclockwise. Satellite images of both sea surface temperature (Color Plate 10a) and ocean color provide spectacular images of Gulf Stream rings. Color differences result from a relative abundance of phytoplankton in the waters to the north of the Gulf Stream and a lack of phytoplankton within and to the south of the Stream. There can be 10 or more rings at a single time, each drifting slowly westward for an average of about 4½ months. Eventually rings coalesce with the parent Gulf Stream and disappear from view. From a ship, the color of the water, measurements of temperature and salinity, or collections of marine life can be used to locate the Gulf Stream and its rings.

The Gulf Stream also has an important influence on climate. It is a great transporter of heat from the tropics to the poles and brings warmth to coastal lands on the East Coast of the United States and along the western shores of Europe. Tropical fish have even been found along the shores of Cape Cod, carried off-course by the Gulf Stream. To the north, where its warm waters collide with the cold waters of the Labrador current, thick banks of fog hover over the sea and land. Turbulence and mixing between the two currents once produced what was one of the most productive commercial fishing grounds in the world, the Grand Banks. The Gulf Stream's northward transport of warm, salty water also facilitates deep-water formation in the North Atlantic and the production of snow and ice in the north's higher latitudes.

Though we know a great deal more about the Gulf Stream than those who first charted its course, there is still much to learn. Scientists today continue to study the Gulf Stream hoping to determine why it meanders and forms rings, what role these rings play in ocean mixing, how flow has changed over time, and how much water and heat are actually transported from low to high latitudes (critical information for climate modeling).

Water's Waves

Wind blowing over the ocean's surface creates not only currents but also waves—a rolling and cresting of the sea that tells of its great power, endless energy, and a changing nature. There are few other ocean phenomena so familiar to so many. Every person who works or plays along the shore or on the sea is well versed with waves and their fatal end at the shoreline, the surf. In the absence of waves, the ocean can appear flat and glossy, like the face of a lake on a still summer day. But when they appear, the sea can roll with huge mountainous swells or signal the approach of a storm in a chaotic, steep chop. Waves are the fortune of surfers, the bane of boaters, and the wreakers of havoc on a sandy beach. Although waves appear to move water toward the coast, in reality they move it up and down, with little forward progression. Waves are simply evidence of the great flow of energy that is eternally transferred through the sea toward the shore.

There are actually two forces that create the up-and-down motion of a wave: a disturbing force and a restoring force. Disturbing forces include wind, earthquakes, landslides, asteroid impacts, changes in atmospheric pressure, and the mixing of fluids with different densities. Restoring forces are gravity and the water's surface tension. When you throw a pebble into a lake or pool, a ring of small waves spreads or propagates outward. The disturbing force is the pebble, and when it hits the water, the energy from its motion (which you imparted by throwing it) is transferred from the rock to the water. Waves are created as a restoring force; either surface tension or gravity, depending on the size of the stone thrown, tries to bring the water back to a level surface.

Disturbing forces push or pull water into a pile, large or small. The water that creates this pile, a crest, comes from the neighboring patch of water. As the crest or pile rises, the surface of the adjacent water lowers into a trough. A restoring force such as gravity then acts on the pile, urging it downward toward a level surface. But because of inertia, the falling pile overshoots its original level and forms a new dip or trough. The falling pile pushes water into the adjacent trough and it rises into a new peak. Adjacent crests successively become troughs, troughs become crests, and a wave moves through the water.

If you were to follow the motion of a parcel of water throughout the passage of a wave, it would go upward as the crest approaches, forward as it passes, downward into the trough, and backward as the trough goes by (Figure 24). This orbital motion of the water creates the wave and its illusion of forward movement. Only if the crest of a wave is higher than the trough is low does the water actually move slightly forward; otherwise it just goes round and round.

Wind is the most common creator of waves in the sea. Whereas a pebble thrown into the water transfers its energy by pushing the water

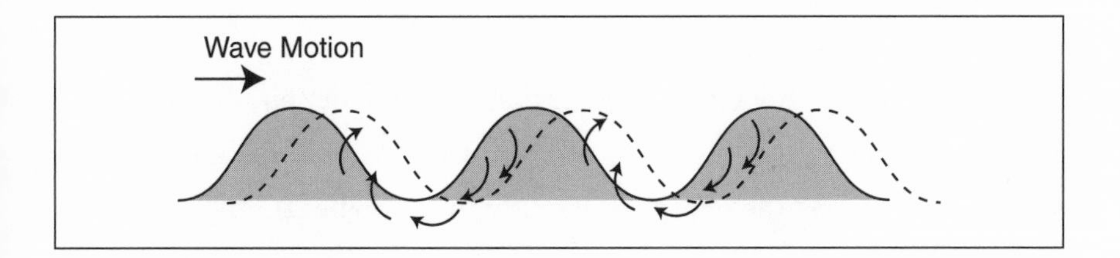

Figure 24 Orbital motion of a wave.

aside, the wind transfers energy by pulling or dragging the ocean's surface—like a flag blowing in the wind, its surface rippling in the breeze. At first, wind blowing over the sea surface generates small ripples. These create a bumpy or uneven surface and make it easier for the wind to "grip" the water. If the wind continues to blow, the ripples grow and gradually form larger waves. At first, they are short and choppy and may seem to come from all directions; this is called *sea*. As waves spread away from the original area of generation they become rolling mounds, called *swell*. Swell forms because longer waves travel faster than shorter ones. Consequently, waves moving away from a storm sort themselves out into a first group of fast, long waves and a later group of slow, short waves. Once the waves reach the coast it is possible to determine how far away the original storm center was by judging the distance between the groups of long and short waves.

The height of a wave in the open ocean depends on the strength and duration of the wind, the depth of the water, and the fetch, or area over which the wind is blowing. In general, stronger winds blowing over a longer fetch will produce higher waves. When waves enter shallow water, the orbital motion of the water near the bottom is flattened by friction from the seafloor and three things happen: the length of the wave shortens, it slows, and then it becomes higher. Conceptually, imagine the bottom portion of the wave slowing and the top continuing at the same speed. When this happens, a wave "bunches up" or becomes steeper, the top overrides the bottom, and it breaks. Essentially, the wave trips over its own feet.

The shape of a wave breaking on the shoreline depends on the wave height, its length, and the slope of the beach. A wave breaking on the shore also transfers the energy of its motion to the beach; the higher the wave, the more the energy. Waves approaching a gently sloping shore tend to spill over and gradually release their energy across the surf zone.

However, on a steeply sloping bottom, waves tend to plunge downward in a magnificent watery curl and rapidly release their energy within a relatively narrow area. If the bottom is really steep, sometimes waves surge forward but do not break because they never get too steep. In the winter and during storms, waves tend to be shorter and higher, so breaking at the shore transfers more energy and can cause extensive beach erosion.

Waves can also reflect off the coast or a seawall, and bend or refract around an obstacle or with changes in depth. Along coasts where depth varies parallel to the shore, waves have an interesting long-term goal: their mission over the millennia is to smooth out the coastline by eroding headlands and filling in embayments. As a line of waves approaches the shore, the portion that moves into shallow water first slows relative to the rest. The part of the wave still in deeper water continues to propagate at the same speed. Hence, the wave appears to bend or curl toward the shallower region and slowing portion of the wave. If you are standing on a cliff or flying low over the shoreline, you can easily observe the bending of the waves around a cliff, pier, or shallow zone. This is called wave refraction; it causes waves to concentrate on shallow regions that stick out from the coast (like a headland) and disperse in deeper areas (like a coastal bay). Consequently, over time waves tend to focus on and erode headlands while spreading out and releasing sediment in bays, thus smoothing out the shoreline.

Rip Currents

Waves approaching the coast can also create rip currents, sometimes mistakenly called rip tides. Swimmers caught in a rip current can get dragged out to sea or drowned when trying to swim against the strong flow. But when one understands what a rip current is, an escape route is unveiled. As wave energy propagates shoreward, a line of incoming peaks and troughs forms parallel to shore. As any surfer knows, the wave height varies across this line, in some areas the wave may be high while in others it is low. As the wave hits the beach, water piles up in the areas where the high peaks hit the shore. Water pushed onshore needs somewhere to go, so it flows from the high to the low areas and then offshore, creating a rip current. The offshore topography of the seafloor often controls where a rip current will form. Rip currents can also be created where obstacles on the beach or in the surf zone funnel water offshore. Although dangerous, rip currents are usually relatively narrow in width. To escape the seaward pull of a rip current and return to the beach, a swimmer should swim to the side or diagonally, not directly against the offshore flowing current.

Waves can also create undertows, particularly on a steep beach, and produce along-shore currents. Undertows are typically caused by the rush of water flowing back to sea after a wave or waves have pushed it onto the shore. If a line of waves hits the beach at an angle, a weak current flowing along the shore is created. Along-shore currents are typically not dangerous, but they play an important role in sand movement down the coast. Currents parallel to shore can cause problems when people construct groins to build up the beachfront. Groins are artificial structures that are built perpendicular to shore to prevent sand from moving along the coast. However, while sand may pile up on one side of a groin, it will undoubtedly erode from the other. One man's beach will grow wide, while his neighbor's will shrink or disappear altogether.

Each year, people spend millions of dollars to combat the ocean and prevent beach erosion, but sediment depletion on a beach is often simply a consequence of the sea's dynamic nature. For instance, during the winter in many regions, relatively large, choppy waves tend to remove sand from the beach and deposit it in an offshore sandbar. Then during the summer, characteristic low, long waves bring the sand back onshore and once again build up the beach. Amazingly, the same beach that disappeared in the winter naturally returns in the summer. Dredging the offshore sandbar or constructing an artificial structure can disrupt this natural cycle and cause a permanent loss of beach sand. We will never completely curtail coastal erosion, as it is a reflection of the ocean's eternal energy. A temporary reprieve is often offered by beach replenishment, the construction of seawalls, sandbars, and vegetated dunes, but eventually whatever sand was once lost, will probably go again. Our best bet is a better understanding of the ocean's actions in the coastal zone and policies for development and recreational facilities that meld with nature, rather than futile attempts to harness the ocean's power and stop the impact of its perpetual motion.

The surf zone is a very difficult and harsh place to work, and instruments (as well as people) placed there tend to corrode and break. Consequently, we have a somewhat limited understanding of the dynamics of wave breaking and sediment movement in the nearshore region. Today's technologies, which allow for smaller and more resistant instruments and sophisticated computer modeling, are beginning to reveal more information about how waves shape the shore.

Tsunamis

Since the birth of the sea on the young Earth, towering walls of water have repeatedly swept through the ocean and crashed onto the shore. Known as a seismic sea wave, or tsunami (not a tidal wave), from the Japanese term meaning "great harbor wave," these colossal waves bring to mind visions of sailors sharing the treetops with sharks and ocean liners stranded on mountain peaks. In 1883, on the island of Krakatoa in Indonesia, the eruption of a supposedly extinct volcano generated a series of great waves, one of which was 41 meters (133 feet) high and raced at 1130 kilometers per hour (700 mph) across the sea. The tsunamis swept over the coasts of Java and Sumatra, destroying 165 settlements and killing 36,000 people. A gunboat, the *Berouw,* was lifted up, carried 2 miles inland, and dropped abruptly. Buildings crumbled, trees snapped like matchsticks, and entire towns were swept away. Throughout history tsunamis have wreaked havoc on the shores of Japan, Alaska, Chile, Greece, Indonesia, Hawaii, and Russia. The Caribbean island of St. Thomas in the U.S. Virgin Islands was hit in 1837, and then again in 1867. New geologic evidence suggests that a massive tsunami struck the Pacific Northwest around 1700. In 1998, a tsunami, estimated at over 15 meters (49 feet) high, crashed onto the northern coast of Papua New Guinea, wiping out entire villages, and killing thousands of people. Just as they have in the past, tsunamis will undoubtedly occur in the future. Our best protection against tsunamis and the potential disaster they harbor is to prepare and warn coastal communities of the risks and impending threats. To do this we need to understand how tsunamis form, travel, and behave once they reach the shore.

Tsunamis can be caused by earthquakes, submarine landslides, asteroid impacts, or volcanic eruptions. Once triggered, they race across the ocean as a series of low, fast waves about 1 meter high, typically traveling at speeds of 800 to 960 kph (500 to 600 mph). At sea, these mountains of water are benign beasts, virtually imperceptible to the human eye. A ship may sit completely unaware as a deadly tsunami passes beneath its hull. The danger lies in wait at the coast.

Imagine blowing over the surface of the liquid in a teacup. Small wind waves are created. If you now shake the teacup, larger waves slosh back and forth. Similar, real-world, sloshing waves are called *seiches* and are often observed in lakes and reservoirs during or immediately after large earthquakes. Tsunamis are commonly generated in a similar fashion, by the violent shaking and deformation of the seafloor.

In wind-generated waves, the orbital motion of the water decreases with depth (with distance away from the wind). At a depth of about one-half the wavelength, the orbital velocity of the water is reduced by 96 percent. Consequently, because energy is transferred through the water's motion, in wind waves traveling through deep water most of the energy is concentrated near the surface. Even for relatively large wind waves, the amount of energy being transferred is limited to the near-surface waters. On the other hand, in a tsunami, the energy imparted to the water during its formation (e.g., an earthquake) sets the entire water column in motion (remember the teacup). Orbital velocities do not decrease significantly with depth, and although the wave height at the surface is relatively small, a few meters at most, the energy contained throughout the entire water column is huge. Furthermore, the rate at which waves lose energy is inversely proportional to their wavelength. Because tsunamis have a very long wavelength, they contain lots of energy, move at high speeds, and travel great distances with little energy loss—it is a prescription for disaster for the coastlines they strike.

Like any other wave approaching the shore, a tsunami entering shallow water begins to feel bottom, causing it to slow, "bunch up," and finally break in a mountainous cascade of water. Tsunamis are often preceded by a leading depression wave that causes a great lowering of sea level as water is sucked up into the growing wall of water. Stories tell of people venturing out to collect fish left stranded by a great retreat of the sea, only to be killed moments later when a tsunami crashes down. One of the highest tsunamis ever documented occurred in 1971 in the Ryukyu Islands of Japan, where a wave reportedly crested at some 85 meters (276 feet) above sea level. The size and force of a tsunami depends on the local topography, shape of the shoreline, and the direction of approach. And an earthquake that generates ripples in one area may cause a tsunami elsewhere. Tsunamis are triggered commonly within and around the Pacific Ocean, where frequent earthquakes and volcanic activity occur.

To prevent catastrophic disasters from tsunamis, scientists and emergency managers are working to establish an effective tsunami warning system throughout the Pacific Ocean. The warning system currently consists of a series of seismometers on the seafloor and moored tide-gauge stations. The seismometers constantly record earthquake activity and seafloor movement, and the tide stations measure changes in sea level. A new addition to the system is presently under development and being tested. It is an instrument that sits on the seafloor and measures changes in pressure as waves pass by overhead (as a wave passes, the overlying pressure increases beneath its crest and decreases below the trough). While the seismometers detect potential tsunami triggering events, the bottom pressure recorder is designed to actually detect a tsunami once formed. If either a triggering event occurs or a tsunami is detected, tsunami warning centers around the world can be notified, the expected landfall determined, and a tsunami watch or warning issued for threatened regions.

In the recent Papua New Guinea event, the warning system was unfortunately of little help. The earthquake occurred so close to shore that the subsequent tsunami hit within minutes. Researchers now believe that a submarine landslide actually triggered the devastating waves, not the earthquake itself, and that this type of tsunami generation may be more common and significant then previously believed. To prevent and predict future incidents like the 1998 event, researchers are using sophisticated computer models and data from recent events to better understand how and why tsunamis occur (see Prager, 1999, for more on tsunamis).

Rogue Waves

Sea captains traveling the waters of the open ocean have recounted amazing tales of colossal waves, some 30 meters high, appearing out of the blue, literally. With no warning, a single huge wall of water can sink or seriously damaged any sea-going vessel. These are called rogue waves and are thought to form when sea conditions allow several individual waves to combine and create one unusually large and destructive wave. Luckily, this seems to be an infrequent occurrence.

Waves also form within the interior of the ocean between layers of differing density; these are called internal waves. The breaking of internal waves on coral reefs and at the sea's edge is now thought to play an important, though poorly understood, role in the mixing of coastal waters, upwelling, and the morphology of the shoreline.

The Rhythm of the Tides

For centuries, humans have noticed the perpetual rise and fall of the sea each day. In some places, sea level oscillates once a day, and in others, twice. The water level may change only a few centimeters or it may vary more than 10 meters (32 feet) each and every day. The largest sea-level variations are created where the shoreline forms a sort of funnel from the open ocean into a restricted embayment. Such is the case in the Bay of Fundy, Nova Scotia, where tides can be an unbelievable 13 meters (43 feet) high. Boats quietly bobbing up and down during high tide lay aground at low tide. Tides can also create fast-moving waves that travel up a river. In the Amazon, the tides combine with a narrowing of the shoreline and changes in depth to create a huge wave 5 meters (16 feet) high that regularly rushes upriver at speeds of over 20 kilometers per hour (12 mph). A similar wave, truly a tidal wave, rushes up the Fu-Ch'un River in northern China at over 25 kilometers per hour and can reach a height of some 8 meters.

Tides also play an important role in the biology of the sea. The distribution of marine animals and plants along the shore is often controlled by the daily ups and downs of the ocean. Other creatures within the sea time their feeding or reproductive cycles based on the tides. During the spring and summer months on California's beaches, a small fish, the grunion, makes its annual appearance to spawn and lay its eggs. During the highest tides the grunion are carried by waves high onto the beach. In an incredible feat of timing, eggs are laid, fertilized, and then buried all just before the tide begins to ebb. The adults are then carried back to sea, while the eggs remain beneath the warm sands to incubate till the next incursion of the sea, the next spring tide. Horseshoe crabs perform a similar feat on the beaches of the East Coast in May and June.

Early on, people sought an explanation for the daily rhythm of the tides. Some suggested that an angel lowering and raising its foot into and out of the sea created the tides. Another early theory was that tides reflected the breathing cycle of a great whale, the inhalation being the ebb and exhalation the flood. But it was Sir Isaac Newton and his theory of gravitation that gave us the first insight into a slightly more scientific explanation.

Newton's theory states that every mass in the universe attracts every other mass with a force proportional to their masses and inversely proportional to the square of the distance between them. In other words, a gravitational attraction exists between all masses in the universe, such as the planets; the greater the masses, the larger the attraction, and the farther away the masses, the smaller the attraction. Both the moon and sun exert a gravitational attraction on Earth. Because water is not firmly attached to Earth, this gravitational attraction acts as a disturbing force and generates a very long wave, the tide. Another disturbing force occurs because Earth and the moon also mutually revolve around a common point, or center of mass. Because Earth is larger than the moon, the common point is closer to Earth, much like an adult and a small child on a seesaw. To balance the seesaw, the adult must sit closer to the center than the child. The force that keeps the moon and Earth in orbit about this point is directed inward and called centripetal acceleration. For simplicity and as a way to visualize its effects, consider an equal and opposite force directed outward, centrifugal force. Think of it as the force that throws you outward against the back of those stomach-wrenching, spinning teacups at an amusement park. Every point on Earth feels an equal centrifugal force away from the center point within the Earth-moon system. It is the balance of these two forces, gravitational attraction and centrifugal force that creates the tides (Figure 25a).

On Earth's surface, points nearest the moon feel a greater gravitational attraction than those on the opposite side, while the centrifugal

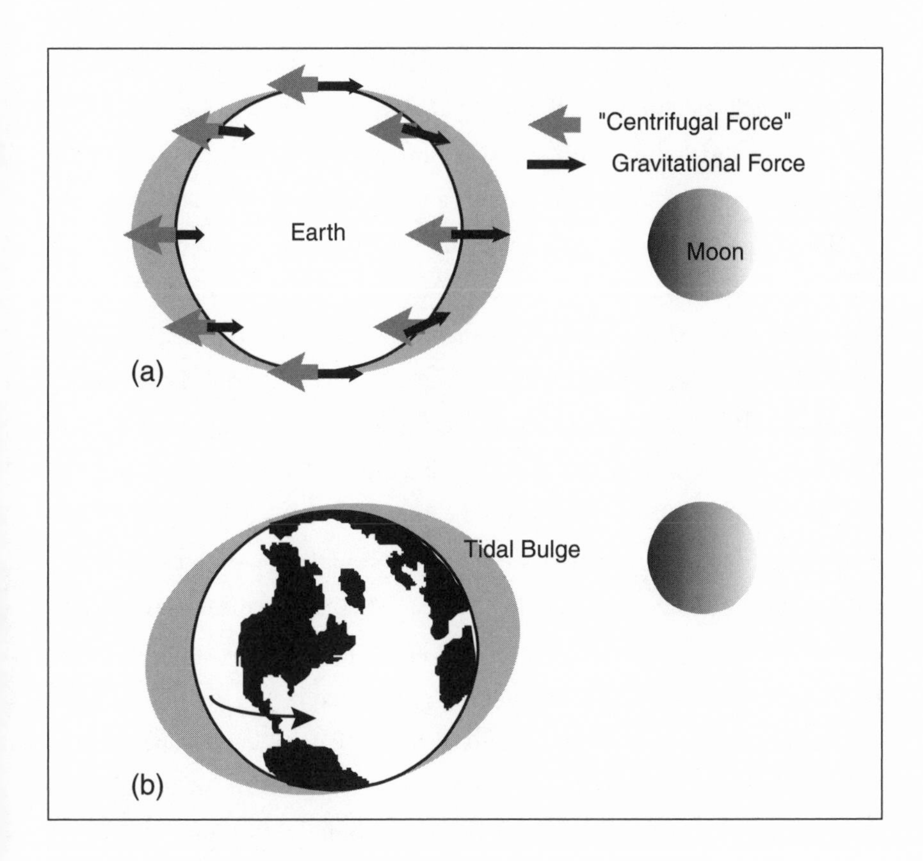

Figure 25 (a) *Tidal forces.* (b) *Tilt due to the moon's position.*

force is equal at all places. The net force at all points creates the tides. If Earth were nonrotating and completely covered with water, it would exhibit an egg-like bulge at its middle, aligned with the moon. If we add the rotation of Earth under the bulges, almost every point on the surface would experience two high tides and two low tides, a semidiurnal cycle. However, the moon is not actually aligned with Earth's equator, but sits at an angle so that the bulges are tilted (Figure 25b). Now one can imagine why on some places there are semidiurnal tides and at other locations there are diurnal tides, one high and one low per day.

Each day at a specific location the tides occur about 1 hour later. The moon rotates around Earth once every 29½ days, so each day the moon moves a little bit more to the east relative to Earth. As Earth spins, it must rotate a little bit past its original starting point to catch up with the moon; this is why the tides are about an hour later every day.

The sun also creates tides on Earth. Although the sun is much larger than the moon, it is also 400 times farther away. Because of its distance, the solar tides are, on average, less than half that of lunar tides. However, when the moon and sun align during the full or new moon, their gravitational effects combine and create the highest of tides, the spring tide. When the moon is near the first or third quarter, at a right angle to the sun, then the smallest of tides, the neap, occur. The cycle of spring and neap tides occurs every two weeks in sync with the moon's rotation. Additional influences on the tides, over 50 components, occur due to other orbital factors such as Earth's tilt about its own axis and the changing distance between the planets during their orbits.

But of course we must return to our real planet where land, changes in ocean depth, and the shape of the coastlines come into play. All of these factors make the tides vary in height and timing from region to region and coast to coast. And because the tides act like a very long wave progressing through the sea, Coriolis once again throws that curve ball. The tides actually move through the ocean basins in a circular sequence, driven to the right in the Northern Hemisphere and to the left in the Southern. Computer models that incorporate all of the lunar and solar tidal components, Coriolis, the shape of the ocean basins, depth, and some baseline measurements, can now predict quite accurately the height and times of the tides throughout the world.

Tides also create tidal currents during their rise (flood) and fall (ebb). Currents are strongest midway through the tide's rise and fall, and are calm or slack at high and low. Where tides enter a bay or estuary, the ebb is often stronger than the flood. An incoming tide can bring large amounts of water into a bay or wetland through a narrow inlet. When the tide reverses, all of this water tries to rush seaward through the narrow inlet, creating an intensified flow. The speed of an outgoing tide can also be enhanced by the seaward flow of a river. In addition, if large storm waves hit the coast during the high or spring tide, flooding and coastal erosion can be unusually severe.

Oceans and Climate

FOR AS LONG as the ocean has existed, there has been an unceasing interplay between climate and the sea. Already we have seen the effects of this interaction on sea level, the extent of ice and snow cover, ocean circulation, global warming or cooling, and the evolution of life. In terms of human populations, the constant dialogue between air and sea can equally create lush gardens, productive crops, and pleasant coastal living or disastrous flooding, severe drought, disease, devastating winds, and blinding dust storms. Climate is generally used to describe the relatively large-scale, long-term nature of the planet's weather. Weather typically refers to shorter-term, smaller-scale, climate-related phenomena, like daily wind, rain, and storm patterns. In general, weather tends to effect Earth on a relatively local basis, while climate impacts it on a global scale. The ocean's influence on climate and weather can be obvious or it can be more subtle in nature. Along the coast, the sea provides warmth in the winter and cooling in the summer. Winters are relatively warm in coastal cities such as Seattle, Washington, or shoreline communities such as Cape Cod, Massachusetts. But in the interior of the continent at a similar latitude, winters can be brutally cold. During the summer in Miami or on the islands of the Caribbean, a cooling sea breeze provides relief from the heat and sun. In contrast, residents of inland Texas and Louisiana face hot, unrelenting heat during the summer months. The sea's surface temperature and evaporative processes also generate and steer some of the world's most powerful storms—hurricanes—and influence global weather patterns through the infamous phenomenon known as El Niño.

Hurricanes

In the majestic temples of ancient Greece, people prayed to Neptune, king of the ocean, for calm seas and smooth waters. With a wave of his trident, Neptune could pacify the sea or whip it into a raging tempest. Today, we look to science to explain the sea's violent behavior, and computer models and weather forecasters to predict it. Hurricanes, once considered clear evidence of Neptune's fury, are one of the most powerful and destructive types of storms on Earth and the product of the interaction between the atmosphere and the ocean. Advances in both oceanography and meteorology over the last several decades have provided great insight into how, when, and where hurricanes form.

When winds exceed speeds of 120 kph (74 mph) a storm becomes a hurricane, a typhoon, or a cyclone. The term hurricane comes from Hunraken, the Mayan god of winds, and is used to describe storms that occur in the Atlantic and the eastern Pacific, off the coasts of California and Mexico. Off northern Australia and in the Indian Ocean, the sea's same swirling tempests are called cyclones, from the Greek word *kykloma* for "coiled serpent." And in the northwestern Pacific, they are called typhoons, from the Chinese phrase *daaih-fung,* meaning "great wind." Although they have different names and occur in different locations, these storms are all born of the same source, the ocean.

The powerful and swirling fury of a hurricane develops only when certain conditions exist in both air and sea. The principal prerequisites are warm water (at least 26°C/79°F), a disturbance of Earth's wind field, and a force to cause wind to spiral (our friend Coriolis). The warmest ocean waters occur within the equatorial tropics. However, hurricanes cannot form right at the equator because Coriolis is negligible. But just to the south and north conditions are ripe, particularly during the hot summer months, so hurricanes tend to form in the summer in two bands about Earth, between 4 and 30 degrees north latitudes and 4 and 30 degrees south latitudes. Storms may form and move past these boundaries, but it is within this narrow band of heat that the sea's tempests are generally born. The disturbance that typically sets the stage for a hurricane to develop is an atmospheric wave in the easterly trade winds.

Every three to four days in the Atlantic during hurricane season (June to November), an easterly wave appears in the trade winds. As the winds blow from Africa to the Americas, the wave creates a low pressure near the surface as air converges and rises to form a crest in the atmosphere. When an easterly wave occurs it may move harmlessly off to the east, but if beneath the crest there exists an extensive region of warm ocean water and lots of warm, moist air, the wave may begin to build into a mounting fury. The underlying warm water must also be deep, extending down at least 60 meters (200 feet); otherwise mixing by the wind will bring cold water to the surface and suck the heat and energy out of the growing storm.

If conditions are right, warm air begins to rise over the ocean and carry moisture evaporated from the underlying warm water into the growing storm. As air rises, more air is sucked in from below and surface winds begin converging toward the low pressure created by the easterly wave. Without anywhere else to go, the converging air rises, picking up more moisture from the hot sea surface. High in the atmosphere, rising moist, warm air is cooled and begins to condense, creating massive clouds and huge thunderstorms. Cloud formation releases vast amounts of energy and causes the density of rising air to decrease, causing it to rise even higher. At the surface, more air then gets sucked upward into a growing column of ascending air and clouds. This causes the low pressure at the surface to drop even further. Converging and rising winds over the ocean then begin to rotate around the central low pressure and form the eye of the storm (Figure 26).

Within the storm's central eye, winds are nearly vertical, so it appears deceptively calm. But on its outskirts—the eye wall—winds are swirling ever faster. In the Northern Hemisphere, winds rotate counterclockwise around the low pressure (think of the wind as flowing into the low and

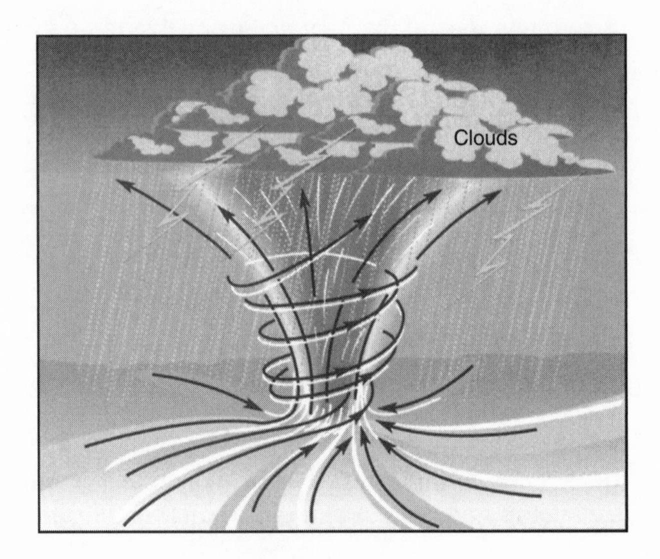

Figure 26 Growth and fueling mechanism for hurricanes. (Reprinted by permission of John Wiley & Sons, Inc. from Exploring Ocean Science, *Keith Stowe, © 1996, John Wiley & Sons, Inc.)*

turning right); in the Southern Hemisphere, they rotate clockwise. At the ocean's warm, moist surface, swirling winds accelerate rapidly and cause even more water to evaporate. Converging winds, heavy with moisture, spin faster and faster and rise within the growing tempest. High above, more thunderstorms are created and even more air is sucked in below. Storm winds reach hurricane strength and continue growing as long as the underlying warm ocean fuels the storm with water vapor. If shearing winds are present high aloft in the upper atmosphere, they can break up a growing storm; or an approaching front can steer the storm over cooler water. But, if conditions have been just right and there are no upper-level winds to break up the storm or steer it away, a hurricane may form, intensify, and move ominously toward land.

In 1998, Hurricane Mitch (Color Plate 10b) struck a powerful blow to Central America; particularly Honduras and Nicaragua. Over 10,000 lives were lost and damage has been estimated in the billions of dollars. The storm's impact was greatly enhanced because it slowed, stalled, and hovered over the region for many hours, dumping enormous amounts of rain. Under the intense rainfall, hillsides cleared for development and agriculture collapsed and became swift rivers of mud that buried thousands of homes and people.

Hurricane Andrew, which hit southern Florida in August 1992, also caused billions of dollars in damage and left thousands homeless, but fortunately due to a timely warning and evacuation, only 43 lives were lost. In this powerful storm, damage was not from rain—only 17.8 centimeters (7 inches) fell on Florida—but its strong swirling winds and downdrafts. Luckily, Hurricane Andrew was moving quickly, some 32 kph (20 mph). Unluckily, it was packing winds of over 200 kph (124 mph) and a storm surge of 2 to 5 meters (7 to 16 feet).

One of the most dangerous aspects of a hurricane and other storms striking the coast is *surge*. Storm surge occurs when the sea rises, rushes shoreward, and then pours off the land. Extensive flooding, high waves, dangerous currents, and widespread coastal erosion can be expected when the sea surges forth. Surge can result from several oceanic and atmospheric conditions. As a storm moves toward the coast, its low atmospheric pressure literally sucks water upward, creating a rise in sea level. More water is driven into the growing pile of water by converging winds at the storm's center. Water may begin to flow out of the pile and surge dangerously toward shore. Strong winds also create large waves that crash onto the shore and elevate the sea even further. If the pressure-induced rise of sea level and strong winds combine with a high tide, the consequences can be disastrous. Storm surge is most dangerous when the storm passes, winds spiral away from shore, and water retreats back to the sea. Swift and hazardous currents erode the land and carry out to sea all that lay in their path. In areas where the seafloor slopes gradually toward shore, bottom friction can enhance storm surge. In 1815, a hurricane struck Long Island with 3-meter high waves atop a 6-meter storm surge (*Restless Earth*, 1997). Nearly 4 meters of water flowed north along the coast, all the way into the streets of Providence, Rhode Island. Moviegoers watching a matinee climbed onto the balcony to escape the onrushing water. Nearly 700 people died and hundreds of homes were destroyed.

Using computer models, oceanographic measurements, and maps of the seafloor and shore, scientists can now predict storm surge. For cities that lie at or below sea level, such as New Orleans, surge predictions are crucial for planning and disaster prevention. One model suggests that if a Category 3 hurricane (see Table 1), struck the northeast United States just right, it could create an 8-meter (26-foot) storm surge in some New York City neighborhoods, with water levels reaching 7 meters (23 feet) near Kennedy Airport and 6 meters (20 feet) in the harbor (*Restless Earth*, 1997).

Table 1: Saffir-Simpson Hurricane Scale (Courtesy of NOAA)

Category	Damage	Central Pressure (millibars)	Wind Speed (mph)	Storm Surge (feet)
1	Minimal	≥ 980	74–95	4–5
2	Moderate	979–965	96–110	6–8
3	Extensive	964–945	111–130	9–12
4	Extreme	944–920	131–155	13–18
5	Catastrophic	≤ 920	≥ 155	≥ 18

Today we have sophisticated satellite imaging techniques, airborne reconnaissance planes, and computer models to help predict the development and track of hurricanes. Satellite images are used to identify and locate a developing storm. To get a more accurate picture of the storm's intensity, the U.S. Air Force Reserve and National Oceanic and Atmospheric Administration fly a plane directly into its mounting fury. The hurricane-hunting plane is specially equipped to measure variables such as wind, pressure, and moisture content within the storm. Storm data are then fed into computer models that also take into account the path of previous hurricanes, sea surface temperatures, and upper-level steering winds. The computer simulations are used to predict the probable paths of the hurricane. Because there are several different hurricane models and they sometimes don't agree, scientists run all of the models and look at what the majority predict before making official forecasts. For the most part, computer models have become fairly accurate; however, nature can still throw in a surprise now and then. Some hurricanes have been known to unexpectedly backtrack, curve, loop, or shoot forward. We still have much to learn about hurricanes so that we can better predict their path and changes in intensity. Recently developed technologies that include improved measurements and three-dimensional imaging of clouds and rain within a hurricane will undoubtedly reveal new insights into hurricane behavior.

On average, only about 10 percent of the easterly waves that form each year develop into full-fledged hurricanes. There is some concern that global warning may increase the number and intensity of hurricanes formed each year. Inevitably, hurricanes will develop, intensify, and move menacingly toward shore. Again and again Earth will unleash its powerful nature on the seas and in the air. As inhabitants of the planet we remain at its mercy, but with a solid understanding of how hurricanes form, move, and strike, we can prepare wisely, warn threatened populations, and try to minimize loss of life, land, and property.

Monsoons

Within the Indian Ocean and its skies above, a seasonal reversal of the winds creates a unique pattern of shifting air and currents, the monsoons. The term monsoon comes from the Arabic word *mausim,* which means "season." The monsoon is in many ways a larger version of a coastal sea-breeze system. During the Northern Hemisphere summer, the land in Asia and Africa heats up. Warm air rising over the land draws in air from the Indian Ocean and creates surface winds and ocean currents that flow to the north and east. A clockwise ocean gyre is thereby produced and winds pick up moisture as they blow across the warm sea surface toward land. Torrential downpours, known as monsoon rains, fall over Asia and North Africa, bringing welcome relief from the heat and water to thirsty crops. Rainfall during the southwest monsoon is not continuous, but tends to occur in short intense bursts that are followed by 20- to 30-day minidroughts, or monsoon breaks.

During the Northern Hemisphere winter, the land cools much more rapidly than the sea, so the system reverses. Air rises over the relatively warm ocean and is drawn in from the continents. Winds and ocean currents at the surface reverse and flow to the south and west, creating a counterclockwise gyre. Moist air is now driven south past the equator to southern Africa. The impact of this reversal of wind-and-water flow off eastern Africa is dramatic. During the summer monsoon a swift, narrow western boundary current (the Somali Current) flows northward along the shore, and coastal upwelling within the region creates fertile waters for fishing. However, come the fall and winter reversal, the Somali Current switches direction and weakens, and coastal upwelling ceases. One of the greatest influences on the strength of the wind and intensity of rain during the monsoon season is El Niño.

El Niño

El Niño may be the best and most dramatic illustration of ocean-atmosphere interaction, and its powerful influence on everyday life on Earth. In 1982 and 1983, and then again in 1997 and 1998, strong El Niño events struck and wrought havoc throughout the globe. Extreme heat and severe drought plagued parts of Indonesia, Asia, Australia, and Africa. Forest fires burned out of control in areas such as Florida, Indonesia, and southern Australia. West coast communities along the United States and Peru were deluged by heavy rains, flooding, massive mudslides, and severe shoreline erosion. The pacific islands of Hawaii and Tahiti were clobbered by powerful typhoons, and hurricanes ravaged the Pacific coast of Mexico. High ocean temperatures caused coral reefs to bleach white and coastal upwelling ceased, leaving fish hungry or causing them to migrate away from their usual haunts. In response, seabirds and sea lions suffered from starvation. Warm weather, rain, and the lush growth of vegetation also spawned high insect and rodent populations. Incidences of both insect- and water-borne diseases such as typhoid, cholera, dysentery, malaria, and yellow fever increased. Such widespread and devastating impacts brought huge attention to El Niño and growing concern over future events. During the 1997–1998 episode, El Niño received so much media attention that it became a household word, the brunt of many late-night jokes, and the scapegoat for many, if not all, weather-related woes. Though scientists and fisherman have long recognized the symptoms of El Niño, 1997–1998 was the first time that researchers were able to accurately predict its onset and effectively warn of its potential consequences.

For many years off the shores of Peru, fishermen have noticed a periodic warming of the coastal waters and a weakening of the normally fertile anchovy fishery. Because it happens around Christmastime, they named this phenomenon El Niño, for the Christ child. When the waters become extremely cold, they called it La Niña, meaning "the girl child."

During an El Niño event, two major changes occur in and over the equatorial Pacific Ocean, one in the atmosphere and one in the ocean. However, much like the old chicken-and-egg conundrum, it is a bit unclear which comes first.

In the 1920s, a British scientist, Sir Gilbert Walker, was trying to find a way to predict the Asian monsoon, and in particular, why some summer monsoons tended to be weak. As he sorted through weather reports from all over the world, he found a remarkable connection in the atmospheric pressure readings between the east and west sides of the Pacific Ocean. When the pressure rose in the east, it fell in the west, and vice versa. Walker called this seesawing of atmospheric pressures in the Pacific the Southern Oscillation. Later, in the 1960s, University of California professor Jacob Bjerknes realized that the unusually warm sea surface temperatures, weak trade winds, and heavy rainfall associated with El Niño were connected to Walker's Southern Oscillation. Hence, this ocean-atmosphere phenomenon is often called the El Niño Southern Oscillation, or ENSO.

During "normal" years, atmospheric pressure is high and sea level is low on the eastern side of the Pacific, off the west coasts of Central and South America. On the opposite side of the Pacific Ocean, near Indonesia and northern Australia, atmospheric pressure is low and sea level high (Figure 27a). Along the equator, the atmospheric pressure gradient helps to drive the trade winds from east to west. Equatorial upwelling brings cold, nutrient-rich waters upward from below and currents flowing from Antarctica along the west coast of South America bathe the eastern Pacific in chilly water. Although the Galapagos Islands lie near the equator, the surrounding waters are cold enough for sea lions, fur seals, and even penguins to coexist with coral reefs and tropical fish—a very

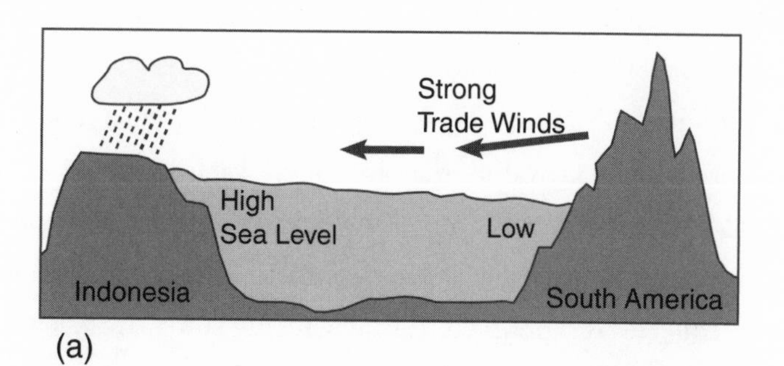

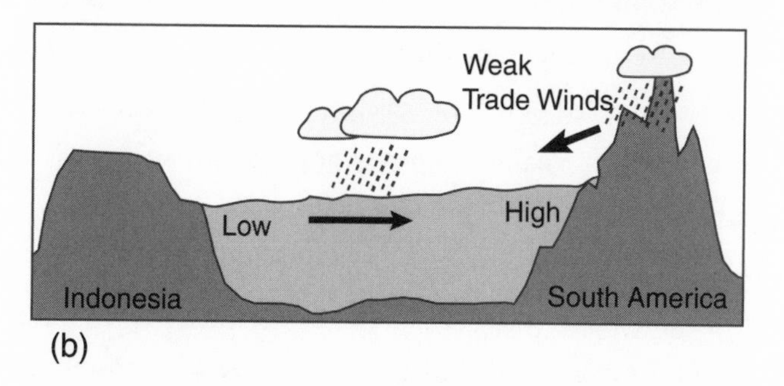

Figure 27 Diagrammatic sketch of (a) "normal" and (b) El Niño conditions in the equatorial Pacific.

strange mix of marine organisms indeed. Normally, coastal upwelling off Peru supports one of the largest anchovy and sardine fisheries in the world and the climate on nearby landmasses tends to be dry and cool. On the western side of the Pacific, near Indonesia and Australia, currents flowing along the equator are warmed by the tropical heat and sun, and bathe the region in warm water. The pool of warm water surrounding the region causes hot air, laden with moisture, to rise. Clouds form and rains normally douse the islands of the western Pacific, stimulating lush growth of vegetation and fertile soils.

El Niño occurs when atmospheric pressures flip-flop, the trade winds relax, and a wave of warm water flows eastward across the equatorial Pacific (Figure 27b). Atmospheric pressure becomes high in the west and low in the east, and a wave of warmth spreads eastward across the equatorial sea. Sea level falls in the western Pacific and rises in the east (Color Plate 8b). Regions like the Galapagos Islands, coastal Peru, and southern California are bathed in waters that are much warmer than usual. In the 1982–1983 El Niño, the warming of the ocean killed 95 to 98 percent of the coral in the waters of the Galapagos. Coastal and equatorial upwelling shut down, fisheries disappeared, and marine life suffered. In 1998, thousands of sea lions off the shore of Peru and California, particularly young pups, died because of a lack of fishy food. Warm seawater temperatures and increased incidences of marine disease are also thought to have played a role in the numerous whales, dolphins, and manatees that washed ashore dead or dying in 1997.

During El Niño years, warm air, heavy with moisture, rises not over the western Pacific but in the central and eastern Pacific. Torrential rains, mudslides, and landslides plague the west coasts of North and South America. Meanwhile, in Indonesia and other areas of the western Pacific, less rain falls and drought as well as forest fires ravage the region. El Niño also weakens the southwest monsoons over southern Asia and influences the frequency, intensity, and paths of major storms. Typically, during El Niño years, fewer hurricanes occur in the Atlantic, while more numerous and intense storms form in the Pacific. El Niño can also modify weather patterns outside of the tropics by shifting the path of the jet stream.

The warming of the eastern Pacific during El Niño further weakens the trade winds and lowers the pressure in the overlying atmosphere. In fact, some believe that changes in ocean temperature across the Pacific initiate an ENSO (versus atmospheric pressure) by causing atmospheric pressures to flip-flop. Because the atmosphere tends to respond to changes in the ocean more quickly than the ocean reacts to changes in the atmosphere, scientists believe that a long-term fluctuation in the sea is what switches El Niño on and off. But the causative agent is far from clear. Some speculate that a long, slow wave may travel counterclockwise around the edge of the Pacific Ocean basin. Every 4 to 7 years it reaches the western equatorial Pacific and begins an ENSO event.

In 1997, scientists were able to forecast the onset of ENSO about 6 months ahead of time. Working with the emergency management community and others, they began to educate people and warn communities about its potential impact. Based on sea surface temperature records and reports of rainfall and atmospheric pressures, fish catches, tree rings, seafloor sediments, and cores taken from corals, researchers now think that El Niño has been occurring for hundreds, if not thousands, of years. But if El Niño has been occurring for so long, why was 1997 the first time scientists could accurately predict it?

Following the 1982–1983 event, the international science community undertook a focused effort to study El Niño and predict the next episode. To do this, an extensive system of observation stations and drifters were deployed throughout the tropical Pacific. Together with measurements of the ocean and atmosphere from ships and satellites, the Pacific ocean observing system provided unprecedented information on the sea and atmosphere. With the tools in place and the clues coming in, scientists recognized early in 1997 evidence indicative of an impending El Niño. Using sophisticated computer models, scientists then simulated changes in the ocean and atmosphere consistent with the oncoming El Niño, illustrated its progression, and predicted its effects. They were also able to monitor and watch as El Niño reversed and La Niña developed.

La Niña is less well understood than El Niño, but it can be just as devastating. La Niña is essentially the opposite of El Niño, characterized by unusually cold waters throughout the equatorial Pacific. Following the 1997–1998 El Niño, Pacific sea surface temperatures began to fall and continued dropping into January 1999. By then, the effects of La Niña were beginning to impact the global climate. The jet stream over North America shifted with the changing atmospheric pressures and air currents over the Pacific. Extreme winter weather struck the northern United States, burying Buffalo, New York, in some 1.5 meters (5 feet) of snow. In just one day, Chicago received a record-breaking 0.5 meters (1.6 feet) of the white stuff. All-time temperature lows hit Indiana and Maine, with frigid readings of −36°F and −55°F. With La Niña conditions in full swing, unusually warm air in the south collided with extraordinarily cold air in the north, setting up conditions ripe for severe warm-weather events. On January 17, 1999, some 32 tornadoes went on a rampage across Tennessee, Arkansas, and Mississippi, and later in the month, another 52 wreaked havoc through the south. Forecasters called for La Niña to weaken by June, but in the meantime abnormally cold and wet weather continued in the Northwest, and warm, dry conditions prevailed in the central and southern United States.

Scientific research has recently revealed that a climate variation similar to El Niño also exists in the North Atlantic. Records of sea surface temperature in the North Atlantic show periods of warm and cold that vary approximately every 10 years (Deser, 1996). Variation in the strength of wind blowing over the warming and cooling sea also shows a similar decadal periodicity. These changes have now been linked to variations in atmospheric pressures over the region and shifts in the climates of both the northeastern United States and northern Europe. During "normal" periods, an area of atmospheric high pressure sits over the Azores, and low pressure sits over Iceland. Between the two pressure centers, wind blows from North America toward Europe, while north of the Icelandic low, and to the south of the Azorian high the winds blow in the opposite direction, to the east (McCartney, 1996). Periodically, the strength and position of the pressure centers oscillates, bringing changes in the surface winds, sea surface temperatures, and climate. Normally, strong winds are heated over the ocean and bring warm air to northern Europe. When the pressure systems oscillate, the Icelandic low moves to the south off Newfoundland and high pressure sits over northern Greenland. Cold, dry polar air then blows across to northern Europe, bringing cooler summers and more severe winters. Milder conditions occur in the northeastern United States as do more nor'easters (McCartney, 1996). Scientists today are more closely studying the 10-year cycle in the North Atlantic and now call it the North Atlantic Oscillation, or NAO. New research shows that the NAO is related to climate oscillations over the Arctic.

Although we can now predict the onset of El Niño using a Pacific Ocean observing system, satellite technology, and computer simulations, the underlying cause of ENSO and NAO remains a mystery. El Niño events seem to occur every 2 to 7 years, but sometimes sooner and sometimes later. No two El Niño events are alike, and some years are more severe than others are. Modern research is quickly revealing that in addition to long-term variation (such as the glacial and interglacial changes), Earth's climate commonly undergoes much shorter, smaller-scale changes. We still do not completely understand how, why, or when many of these variations occur or what their impacts will be.

Because of the attention brought to ocean observations through phenomena such as El Niño and hurricanes, many hope that a global ocean observing system will soon be developed and implemented. The system would set up an extensive series of worldwide monitoring stations and drifters, combined with satellite imagery, computer models, and other technology to constantly observe the oceans and provide daily information, much like what we see today with the weather. Observations would take place in the open ocean, within the ocean's interior, at the seafloor, and along the coast. A global ocean observing system could provide information for a vast number of important purposes, including better forecasting of climate and weather patterns, improved safety and efficiency of marine operations, a better understanding of inputs to the sea and marine ecosystems, and data that would help us to understand the occurrence of harmful algae blooms and the spread of ocean-associated disease. A global ocean observing system would also enhance our understanding of two other phenomena of great interest today—global warming and sea-level change.

Global Warming

Throughout Earth's 4.5-billion-year history the interactions of air, sea, and land have caused the climate to fluctuate between periods of warmth and cold. The ocean has always played a major, though somewhat murky, role in the planet's changing climate. Today, great concern and political debate centers around Earth's shifting temperatures—more specifically, its heating up at an "unnaturally" accelerated rate, a phenomenon known as global warming. Global warming itself is not the greatest cause for alarm, but the rapid pace at which it is occurring is. When climate change has occurred over time periods of thousands, hundreds of thousands, and millions of years, Earth's creatures had time to adapt, migrate, and alter their lifestyles as their surroundings changed. On the other hand, when climate change occurred more rapidly, many of the planet's inhabitants were wiped out. We are creatures of Earth, and an unnaturally fast heating of the planet threatens our health, quality of life, and possibly our very existence. Extreme heat, sea-level rise, flooding, disease, drought, and intensified storm activity have all been cited as coming consequences of global warming. Though there has been extensive debate about global warming, its cause, rate, and impacts, it is now widely accepted that the planet is heating up and we are partially to blame.

In 1997, the global average surface temperature was the warmest of the century and possibly of the past 1000 years; during 1998, globally averaged surface temperatures reached an all-time high each and every month. A 1995 report on climate change by the Intergovernmental Panel on Climate Change (IPCC) suggests that surface temperatures this century are as warm as, or warmer than, any other century since 1400 A.D., and worldwide, the average surface temperature has risen by 0.3 to 0.6°C (0.5 to 1°F). As a result, sea level has risen 10 to 25 centimeters (4 to 10 inches) and glaciers are melting. It is also clear that although many factors can contribute to global warming, increasing levels of carbon dioxide in the atmosphere due to fossil fuel burning and deforestation are rapidly causing Earth's thermometer to rise.

Carbon dioxide, water vapor, and the other greenhouse gases (methane, nitrous oxides, chloroflourocarbons, and ozone) absorb longwave, or infrared, radiation emitted by Earth. Absorption of this radiation heats up the atmosphere and warms the planet's climate. The composition of air bubbles trapped in ice cores from the Antarctic, and measurements of the atmosphere at Hawaii's Mauna Loa Observatory, indicate that since 1850, the amount of carbon dioxide in the atmosphere has increased by about 25 to 30 percent. So the question becomes, with increasing levels of carbon dioxide, just how warm will the planet get and at what rate? The IPCC report suggests that Earth's average surface temperature will rise by 1 to 3.5°C (about 2 to 6.5°F) and sea level will increase by 15 to 95 centimeters (6 to 38 inches) by the year 2100. Where do these numbers come from, and why is there such a wide range or uncertainty in predicted temperature and sea-level rise?

To estimate the rate at which the global thermometer has risen we examine past records of temperature change. Documented air and sea temperatures are used in conjunction with estimates based on the isotopic composition of foraminifera in deep-sea cores, from glacial ice or snow, and in coral skeletons. To predict what the climate will be like in the future, scientists must rely on sophisticated computer models. These models use mathematical equations to represent physical processes and interactions in the atmosphere, ocean, and on land. A starting point is usually based on current measurements or estimates of past conditions. Then, using a spherical grid laid out over the entire globe, thousands of calculations are performed at grid intersections to represent and assess how conditions in the air, in the sea, and on land will change over time. Because of their complexity and size, supercomputers must be used to run full-scale climate models. Much of the uncertainty in their outputs come from the way that various aspects of the climate are represented by different models, and even more so, because there are aspects of climate that are not well understood—one of which is the ocean. In fact, because we do not fully understand how the ocean influences climate, it has been referred to as the wild card in computer simulations of climate change (Karl, 1993).

The ocean's role in global warming stems principally from its huge capacity to absorb carbon dioxide and to store and transport heat. In the sea, photosynthesis by marine plants and algae, principally phytoplankton, removes great quantities of carbon dioxide from the atmosphere. Hence, the greater the growth (productivity) of phytoplankton in the sea, the greater the removal of carbon dioxide. But what controls the ocean's productivity? There are several limiting factors that will be discussed later, but results from a recent experiment suggest that in areas of the ocean where other nutrients are plentiful, iron may be one of the most important and, until recently, unrecognized variables controlling phytoplankton production. Scientist John Martin, who originally proposed and later proved the influence of iron on phytoplankton growth in the Southern Ocean, was known to have half-jokingly said, "Give me a half-tanker of iron and I'll give you an ice age" (Chisholm, 1992). In line with Martin's thinking, some have proposed a radical, highly controversial, and uncertain means to counteract global warming—seeding the oceans with iron to induce phytoplankton blooms. Perhaps increased phytoplankton growth would use up a significant amount of carbon dioxide in the atmosphere, but perhaps not, and there might well be side effects that could be detrimental to the ocean ecosystem.

Within the ocean the production of limestone, in the form calcium carbonate skeletons or shells, also reduces atmospheric carbon dioxide. However when deposits of limestone become exposed and weathered on land or are recycled at a subduction zone, carbon dioxide is released back into the atmosphere. What is not well understood is how much carbon dioxide resides in the sea and at what rate it is taken up and recycled. Relatively new research has also discovered beneath the sea a new and potentially significant threat to skyrocketing Earth temperatures: gas hydrates. Gas hydrates are a solid, crystalline form of water, like ice, except that they contain additional gas, typically methane, and are often found stored in ocean sediments. Large hydrate accumulations have been found undersea off the shores of North and South Carolina, and in the Gulf of Mexico. Increased ocean temperatures could cause gas hydrates to dissociate, releasing massive amounts of methane gas into the atmosphere and cause undersea landslides in the process. Consequently, hydrates may, if released, significantly increase global warming as well as create a geologic hazard to offshore drilling operations.

The ocean is also a great reservoir and transporter of heat. Heat from the ocean warms the atmosphere and fuels tropical storms. Heat is transported by currents from the equator to the poles. Ocean circulation, as described earlier, is strongly controlled by wind and the sea's balance of salt and heat. Scientists think that climate warming may slow down circulation, while cooling may speed it up, but these responses are not well understood. Evaporation from the ocean also supplies the precipitation that creates fields of snow and ice at high latitudes. Snow and ice coverage change the reflectivity of Earth's surface and are an important influence on how much incoming radiation is either absorbed or reflected. Furthermore, clouds and water vapor in the atmosphere come mainly from the sea and strongly influence climate. Surprisingly, clouds are one of the least understood and poorly modeled parts of the climate change equation. The resolution of most grids used in climate modeling is too large to take into account common-sized cloud formations. Aerosols, tiny particles of soot, dust, and other materials, are thought to seed cloud formation, scatter incoming radiation and promote cooling, but this effect, which would counteract warming, is also only superficially understood. Computer models of climate change must take into account all of the processes within the ocean, over land, and in the sky that potentially influence warming. No wonder there is such uncertainty. Volcanic eruptions that spew dust and ash skyward can also block out the incoming radiation and cool the climate. In 1991, the massive eruption of Mount Pinatubo in the Philippines ejected 20 million tons of sulfur dioxide to heights of 25 kilometers. The eruption is thought to have caused the abnormally cool summer of 1992. How would one know where and when to add a Mount Pinatubo–sized eruption into a climate model? Models must also take into account ocean and temperature variations associated with phenomena such as ENSO, the decadal scale changes in the Atlantic Ocean, and the orbital variations thought to control glacial and interglacial episodes.

More research and improved monitoring is required to better understand interactions within the ocean-atmosphere-land system we call Earth and to predict the future in terms of global warming. Although there is much uncertainty associated with global warming, several things are clear: the planet is warming, carbon dioxide in the atmosphere is increasing, and humans are contributing to the carbon dioxide buildup. Two main issues are now at hand: Should we act now to reduce anthropogenic global warming? If so, what should be done? The answers to these questions appear logical and obvious, but economic concerns and politics blur the issue. Nevertheless, based on what we do know, the time to act is now. Carbon dioxide has a long-lasting influence on the atmosphere, so efforts to reduce atmospheric levels must start sooner rather than later if benefits are to be gained relatively quickly. We know from the pages of Earth's own history that rapid climate change can have an irreversible and catastrophic impact on the planet's inhabitants. Do we want to be the cause of the next mass extinction, knowing that humans could be one of the species to go? The United States and the international community—including politicians, scientists, industrialists, environmentalists, and the public—must work together to find economically sound, safe, politically viable, and effective ways to reduce greenhouse gas emissions, conserve, reduce deforestation, and look for new and cleaner ways to supply energy.

The Yo-Yoing of Sea Level

In a synchronized dance of water, air, and land, sea level rises and falls with Earth's climate swings. During periods of warmth, sea level rises from the thermal expansion of seawater and the melting of glacial snow and ice. In times of glacial cold, sea level drops as water is trapped in ice and seawater contracts due to cooling. Today, as the climate warms, we face a rising sea, on the order of 10 to 30 centimeters (4 to 12 inches) per 100 years. Some worry that if global warming continues the West Antarctic ice sheet will become unstable and collapse. If all of its ice were released into the ocean, sea level could rise an estimated 4 to 6 meters (13 to 20 feet) and cause major coastal flooding. If all of the ice in both Antarctica and Greenland were to melt, sea level could rise some 65 to 80 meters (210 to 260 feet). The current shoreline would be completely submerged and inland realty would soon become valuable beachfront property. The regions of the planet inhabitable by humans would be significantly reduced, while marine environments would expand—the fish would be happy. This is not an unrealistic scenario given Earth's past history. For instance, imagine sea level at least 6 meters (20 feet) higher than it is today and the islands that now make up the Florida Keys completely submerged, as they were some 125,000 years ago. On the other hand, if climate were to swing toward another ice age, sea level could drop some 120 meters (395 feet) as it did during the last major glaciation. But even before Earth began its glacial-interglacial dance, the sea rose and fell over time. Factors other than the highs and lows of climate can influence the ocean's volume and, therefore, sea level.

During the Cretaceous period, 140 to 65 million years ago, the climate was exceptionally warm and sea level was high as a result of thermal expansion, intense volcanic activity, and extensive and fast seafloor spreading. Massive volcanic eruptions released huge amounts of carbon dioxide and water vapor into the atmosphere and fueled greenhouse warming. High rates of seafloor spreading, the formation of an early midocean ridge in the Atlantic Ocean, and the spewing of vast quantities of molten material onto the seafloor also elevated sea level. Imagine a container filled with water: the water represents the sea, the container is the ocean basin, and the edges are the shoreline. If a layer of sand or hot molasses is dumped into the water-filled container, the incoming material displaces some of the water and it spills over the container's edges, inundating the imaginary shoreline. The same thing happens when the seafloor spreads rapidly or massive volcanic eruptions exude vast quantities of molten rock beneath the sea. Seawater is displaced upward and the ocean rises relative to the land.

Orbital variations that affect the amount of incoming solar radiation, massive mountain-building events, changes in the reflectivity of Earth's surface, and removal of carbon dioxide from the atmosphere by limestone production or photosynthesis can also alter sea level. These influences tend to impact sea level on a worldwide scale, but other factors can affect it on a more local basis. Regional tectonic events, such as an earthquake and the uplifting of land, can cause an apparent fall in local sea level (global sea level is unaffected). On the other hand, subsidence of the land can raise the local sea. Where large rivers carry great quantities of sediments into the ocean, the weight of the sediment often causes the land to sink or subside. At the mouth of the Mississippi River and Chesapeake Bay, subsidence is causing sea level to rise rapidly, and both land and wetlands are being drowned. In some regions, sediment deposition and wetland growth can keep up with this relative sea-level rise, but where dams, river diversion, or construction prevent new sediment from reaching a river's mouth, sinking is accelerated. In Chesapeake Bay, a lighthouse on Sharp's Island, once a harbinger of land for sailors, is now falling into the sea (Figure 28). Here, subsidence is creating an accelerated rate of local sea-level rise, about twice the global average, at some 4 millimeters per year (1.3 feet per 100 years). The melting of glacial ice and snow actually

Figure 28 Lighthouse on Sharp's Island falling into the sea due to subsidence and accelerated sea-level rise at the mouth of Chesapeake Bay.
(Courtesy of Curt Larsen, U.S. Geological Survey.)

has two effects on sea level: on a global scale it can raise sea level; locally it can cause a relative fall in sea level. As a glacier melts, the removal of the ice and snow's weight causes the land to rebound or rise.

As early as the 16th century, Leonardo da Vinci found fossils of marine organisms from the Mediterranean in rocks high above the sea and recognized this as evidence that the sea had once been higher. Scientists today use methods similar to those unwittingly pioneered by da Vinci. Geologists and oceanographers scour the planet looking for the remains of marine organisms or sediments that lay in a position reflecting an earlier level of the sea. Researchers must look carefully and adjust for any vertical movement of the land that might bias interpretations of a past sea-level height. Once an appropriate deposit or outcrop is discovered, scientists typically date accompanying rock or fossils using radiometric techniques. Previously higher sea-level stands are often denoted by coral reefs that now lie stranded above the ocean. Famous examples of these occur on the islands of Barbados, St. Croix, Papua New Guinea, the Bahamas, and the Florida Keys. Rocks once subject to the ocean's erosive forces can also preserve distinctive evidence of a past sea level. Typically, a flat, narrow platform, called a wave-cut terrace, is created by erosion at the sea's edge. Rocks subject to wave erosion can also look like swiss cheese and exhibit an abundance of sea caves. Inland ridges of beach sand and shell also suggest a time and place of higher sea level.

Lower stands of the sea are estimated based on similar types of clues, but instead of looking above the current waterline, we must search below. A wave-cut terrace, evidence of exposure to air, or the bones of land-dwelling species found submerged below the sea can be indicators of a lower sea-level stand. Scientists recently discovered the bones of woolly mammoths, saber-toothed cats, and other long-gone terrestrial creatures 19 meters (60 feet) underwater just 27 kilometers off the shores of Sapelo Island, Georgia. And in the Bahamas, Florida, and elsewhere underwater caves contain a remarkable collection of long stalactites and stalagmites that could only have formed during lowered sea level, when the caves were filled with air rather than water.

Though sea level has gone up and down over the planet's history, the acceleration of sea-level rise from global warming concerns many, especially those living within the most vulnerable areas. Low-lying regions like Bangladesh and coastal communities and cities such as New Orleans, which lie near or below sea level, are most at risk. Rising sea level coupled with global warming may lead to impacts such as increasingly frequent and intense flooding during storms, the spread of water-borne or related disease, loss of property and crops, and the intrusion of saltwater into coastal aquifers—our reservoirs of freshwater. Some believe that global warming may prevent us from entering another ice age. Others think that the greenhouse effect will go on heating the planet and continue, if not accelerate, the rise of the sea.

Color Plate 1
(a) The submersible Alvin.
(Courtesy of OAR/National Undersea Research Program, NOAA.)
(b) Clams clustered around gas seeps in the deep Gulf of Mexico.
(Courtesy of Harry Roberts, Louisiana State University.)

Color Plate 2
Life in the ancient seas: *(a)* Cambrian period. *(b)* and *(c)* Mesozoic era.
(Courtesy of the Smithsonian Institution.)

(2c)

(15b)

(15c)

Color Plate 15
(a) A small red fish sits camouflaged within a reef waiting patiently for a meal. *(b)* A cowfish sculls by; its boxy body and small fins make it a slow, but highly maneuverable creature. *(c)* A sea turtle hides out under a ledge. It is missing a flipper, probably as a result of a shark attack.
(E. Prager.)

Color Plate 16
(a) Humpback whale breaching.
(Courtesy of the National Marine Sanctuary Program, NOAA.)
(b) Sea lions in the Galapagos Islands. An aggressive male barks on a beach near a towering pinnacle of volcanic rock.
(E. Prager.)
(c) Weary after hours of play and feeding, sea lions nap on a beach.
(E. Prager.)

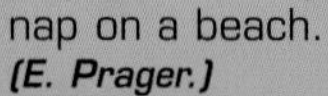

(16b)

The Geological Ocean

As the wind and living sea flow with time, so too does the underlying Earth
—Sylvia Earle

LIKE THE INTERTWINING of the seas themselves, the geological ocean is inseparably linked to the biological, chemical, and physical sea. Wind and currents that transport heat and water also carry particles of rock and organic debris. Within the ocean, these particles rain down in an endless underwater snowfall that covers the seafloor in a blanket of soft sediment. Deceptively biological in nature, coral reefs are truly geological in form, and beneath their thin skin of living tissue lurks a massive accumulation of limestone rock. In the deep sea, ridges, rises, and gullies funnel the ocean's unceasing flow, like a maze guiding a restless wanderer, and circulation through cracks and fissures within the underlying rock alters the chemical makeup of sea water.

Unquestionably, one relatively recent revelation has altered the course of history in geological oceanography and is considered one of the most significant achievements in ocean science during the last half-century. Discovery of the mid-ocean ridges, seafloor spreading, and the synthesis of these findings into a theory known as plate tectonics revolutionized the way we think about Earth, its oceans, and the evolution of marine life. A puzzling set of clues, a trail of scientific ingenuity, and as some would call it, dumb luck, led to the unraveling of one of Earth's greatest stories, carrying plate tectonics past theory and into a state of being.

Plate Tectonics Discovered

The unearthing of plate tectonics, so to speak, really began when people started to seriously question the puzzle-like fit of the continents. Leonardo da Vinci in the 15th century, Francis Bacon and a Dutch mapmaker, Abraham Ortelius, in the 16th century, and naturalists Georges-Louis Buffon and Alexander von Humboldt in the 18th century all considered this geographic enigma. In 1858, geographer Antonio Snider-Pelligrini was one of the first to illustrate on a map how the continents may have once fit together. At the time, it was thought that only an extremely powerful earthquake or flood of biblical proportion could have possibly caused the colossal breakup of the once-fused continents. Then, in the early 1900s, Alfred Wegener, a young German meteorologist, proposed a radical, evidence-based theory. He called it continental drift. Wegener hypothesized that more than 200 million years ago the continents had been part of a much larger supercontinent and that some 100 to 150 million years ago the giant landmass separated. Furthermore, said Wegener, as the continents drifted apart, large ocean basins were formed in between.

Wegener based his theory of continental drift on several powerful pieces of geologic evidence. The presence of similar fossils and rocks on now distant shores strengthened the remarkable fit of the coastlines as an indicator of an ancient supercontinent. In eloquent simplicity, Wegener envisioned it as a newspaper page torn in two: "It is just as if we were to refit the torn pieces of a newspaper by matching their edges, and then check whether the lines of print run smoothly across. If they do, there is nothing left but to conclude that the pieces were in fact joined in this way." For Wegener, fossils and rocks were the print and the continents were the paper. He also discovered that fossils characteristic of one climate were now found in environments having a very different climate. In the hot, arid valleys of Africa he discovered glacial deposits, and in the cold, polar regions he found fossilized ferns indicative of a once-tropical clime. In Antarctica, Wegener also found coal deposits that suggest a previously warm climate and tropical vegetation. He hypothesized that these strange findings could only be explained through continental drift. The climates of these regions must have changed as the continents slowly drifted across Earth's surface, moving through space as well as time. He also noted that in the Nordic region (Norway, Sweden, and Finland), following the melting of ancient glaciers and the removal of the ice's weight, the underlying land rose (a process now called glacial rebound). Wegener surmised that if the land could move vertically, it could also move laterally across Earth's surface and that mountains formed of folded rock could be evidence of this horizontal motion.

Unfortunately for Wegener, he introduced his theory at a time when most scientists staunchly believed that the continents and oceans were fixed features on the surface of Earth. Rollin Chamberlain of the University of Chicago unequivocally stated his reaction to Wegener's hypothesis: "Can geology still call itself a science, when it is possible for such a theory as this to run wild." Another well-known and respected geologist referred to the theory as "utter, damned rot!" Wegener's evidence was largely qualitative, and there was a fatal flaw in his theory. He could not come up with a reasonable explanation for how the continents moved, the driving force. Scientists throughout the world, particularly in the

United States, considered his theory unproven and quite unbelievable. Without better, more definitive evidence, the theory of continental drift remained ignored for more than half a century. But until his death in 1930, during an expedition across the Greenland ice cap, Wegener doggedly pursued evidence to prove his theory of continental drift.

The geologic key to unlocking the mysteries of plate tectonics would come from advances in technology that allowed detailed study of the ocean floor for the first time. It was a breakthrough that has been described as a combination of shrewd analysis and serendipity. Before World War I, technology for studying the seafloor was relatively crude and our understanding of the ocean and its sediments was superficial at best. Depths were measured by soundings using a long line or cable weighted at the end, and sediments were collected by difficult and tedious dredging operations. After World War I, primitive sonar systems were developed to measure depth by bouncing a pulse of sound off the bottom and recording its travel time back to the ship, a technique known as echo-sounding. Using echo-sounding, scientists were able to make detailed bathymetric or depth surveys of the seafloor, and by the 1950s they had discovered the presence of an enormous, globe-encircling mountain chain beneath the sea—the mid-ocean ridges. Though the presence of some sort of topographic high, possibly a plateau, had been discovered in the 1920s during the *Meteor* expedition, it was not until the 1950s that scientists recognized the true enormity and nature of the ridge system. The mid-ocean ridges rise an average of 4500 meters above the seafloor and wrap around the globe for more than 60,000 kilometers (Color Plate 5). Though undoubtedly the most prominent topographic feature on the planet, the mid-ocean ridge system was not even discovered until the mid-20th century!

After World War II, further advances in marine technology and increased interest in the ocean and seafloor led to two more startling discoveries. Using an instrument called a magnetometer, originally developed to detect submarines, scientists discovered odd magnetic variations on the seafloor. When molten rock containing magnetic particles or minerals cools, its magnetism becomes aligned with Earth's magnetic field, much like a compass. A magnetometer measures the direction and angle or declination of a rock's magnetism. Today, the magnetic properties of a rock cooled at the surface indicate an alignment to the north. Its declination depends on the latitude where it cools. In the 1950s, oceanographers working with the U.S. Navy found a zebra-like pattern of magnetic anomalies or variations on the seafloor (Figure 29). In 1963, Fred Vine and Drummond Matthews of the Canadian Geologic Survey proposed that this striped magnetic pattern was produced by repeated reversals of Earth's magnetic field. Geologists studying rocks on land had previously theorized that throughout Earth's history its magnetic field had reversed or flip-flopped a number of times. Today, compasses point to the north when aligned with Earth's magnetic field; this is called normal polarity. Following a reversal of Earth's magnetic field, a compass would point south, thus having a reversed polarity. Scientists today agree that reversals have occurred over 100 times in the past 75 million years, but they are still mystified about how and why it happens. Will boaters navigating by compass someday become lost or run aground because all of the sudden the compass will swing south? Or will the compass start swinging south gradually, and if so, what will initiate the shift? Vine and Matthews suggested that the adjacent stripes on the ocean floor, positive and negative magnetic anomalies (variations), reflected formation during times of normal and reversed polarity, respectively. However, what was truly curious about the magnetic stripes was that they ran parallel to a midocean ridge, and their spacing and width were the same on either side of the ridge's crest.

Just a year earlier, in 1962, Harry Hess of Princeton University, a Wegener fan, speculated that ocean crust is created by volcanism at the

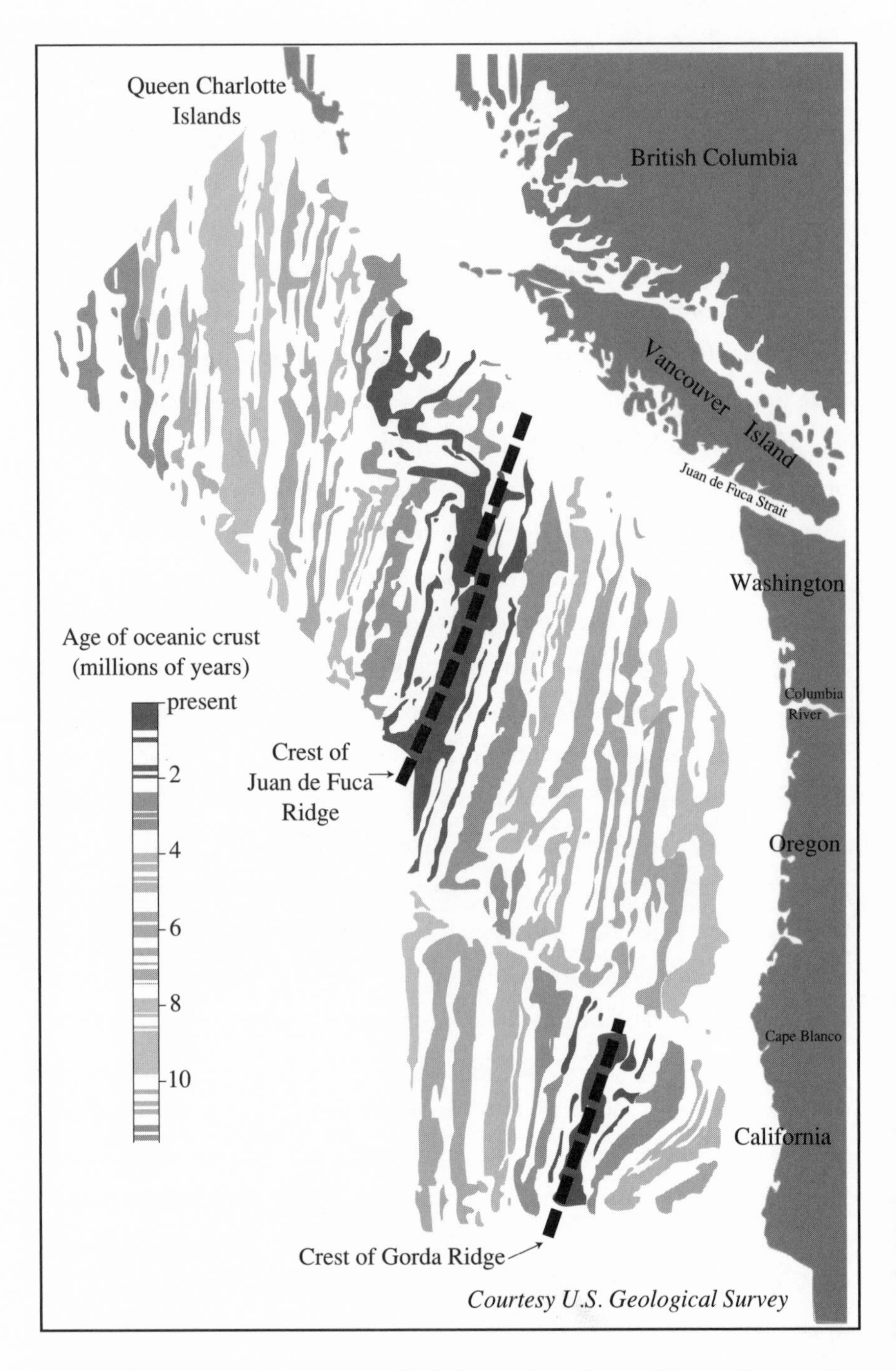

Figure 29 Magnetic variations (shaded stripes) and age of the seafloor in the Pacific Northwest.

crests of midocean ridges, spreads away, and is eventually destroyed in deep ocean trenches. Hess and colleague Robert Dietz called this process *seafloor spreading*. The Vine and Matthews theory for the magnetic striping could be explained by the new concept of seafloor spreading. The ocean floor was a sort of magnetic tape recorder. At a ridge axis, newly forming ocean crust cools and takes on the alignment of the current magnetic field. Spreading of the seafloor then slowly moves the crust away on both sides of the ridge axis. Over time, with repeated reversals of the magnetic field and continued seafloor spreading, a symmetrical, zebra-stripe pattern forms about the ridge axis.

The confirmation of seafloor spreading and its relationship to the magnetic striping was to come during the early days of the Deep Sea Drilling Project (DSDP). Using a specially designed ship (Color Plate 4), and some 6100 meters (20,000 feet) of drill pipe, the DSDP cored and then dated samples of the seafloor. It was an amazing feat, likened to drilling a hole in a New York sidewalk with a strand of spaghetti dangled at night, in swirling winds, from atop the Empire State Building. DSDP data confirmed that with distance away from a ridge axis, the age of the ocean crust increased—the seafloor did indeed form at the axis and spread outward over time (Figure 29). Two other interesting observations supported Hess's radical idea of seafloor spreading. Oceanographers and geologists had previously puzzled over why, although Earth is approximately 4.5 billion years old, only a relatively thin layer of sediment covers the seafloor, and the oldest seafloor fossils are just 180 million years old. In 1965, Canadian scientist J. Tuzo Wilson proposed an idea that allowed for the synthesis of continental drift and seafloor spreading and explained these last two observations.

Wilson, who had been studying earthquakes and faulting in the ocean crust, suggested that Earth's stiff outer rind is broken into a number of moving pieces, or "plates," and that convergence and destruction of these plates at ocean trenches balances spreading at midocean ridges. Fractures across the ridges that he called *transform faults* allowed for motion of the relatively flat plates over Earth's spherical surface. The distribution of earthquakes, and later volcanoes, lent credence to the idea that Earth was divided into numerous platelike sections (most occur along plate boundaries). The destruction of the ocean crust at the trenches could account for the relatively thin sediment cover and young age of the seafloor: the older seafloor and sediments were being recycled back into Earth.

In the late 1960s, new fossil discoveries were also providing evidence in support of the idea that the continents were shifting and moving about Earth's surface. In Antarctica, researchers found fossils of a sheep-sized reptile, *Lystrosaurus,* identical to 200-million-year-old fossils found in both Africa and India; this indicated that at one time Antarctica, Africa, and India were connected.

Throughout the 1960s and 1970s evidence and support for the theory of plate tectonics mounted, and soon its remaining aspects were outlined, including a driving force. The scientific community was given a whole new way of looking at Earth and the forces that create and mold its surface. What was once only a theory became the accepted dogma, and textbooks everywhere had to be rewritten. We no longer call it continental drift or the plate tectonics theory, but simply plate tectonics.

Understanding Plate Tectonics

Before plate tectonics and studies in modern earth science, people generally thought that the planet was solid throughout its interior and fixed on the surface. But Earth is now known to be a dynamic sphere whose hot interior moves in slow motion and surface shifts and changes with time. The planet's hard surface is relatively thin and composed of a series of interlocking rigid pieces, or plates, that move atop a layer of hotter, more fluid material—something akin to a Milky Way candy bar with a hard, thin layer of brittle chocolate overlaying a softer, more fluid caramel below.

Inside the Earth

In 1864, Jules Verne wrote *Journey to the Center of the Earth,* in which his fictional characters embark on a mission to travel into the Earth's depths and unravel its internal mysteries. In reality, we have yet to discover a means to literally view the interior of our planet, so we must rely on indirect methods to observe and study its nature. To create a picture of Earth's internal structure, scientists examine rocks that have been uplifted or ejected from deep within, drill deep cores into the ocean crust, and measure how vibrations, called seismic waves, pass through its interior. The details of the planet's geologic portrait are still evolving, but the following describes what we think Earth's internal composition and physical properties are.

The interior of Earth is composed of several concentric layers (Figure 30). These layers can be divided in two ways—by their chemical composition or their physical properties. Earth is composed of three main layers: an outer crust (0–50 km), an intermediate rocky mantle (50–2900

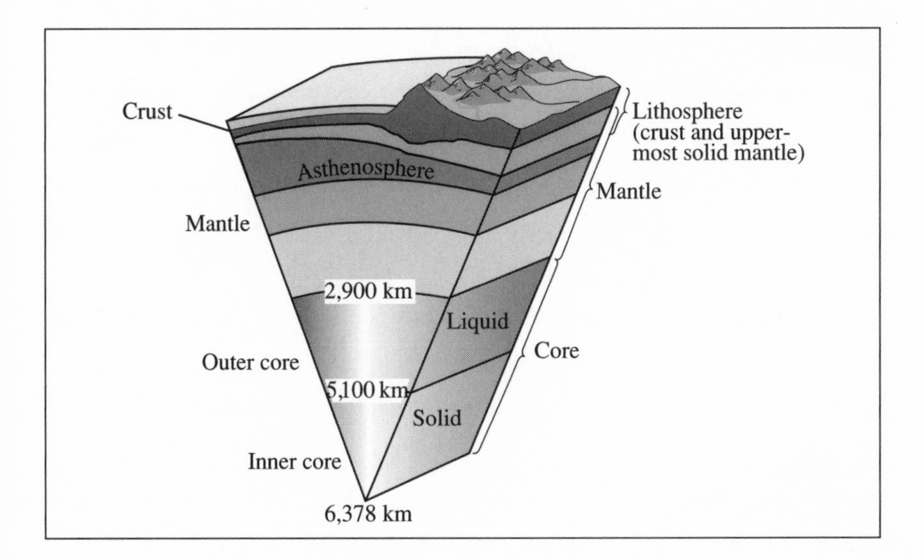

Figure 30 Sketch of Earth's internal concentric layering.
(Courtesy of the U.S. Geological Survey.)

km), and inner metallic core (2900–6378 km). The outermost layer, the *crust,* is relatively thin and includes all that we see on land or beneath the sea. Relative to the rest of Earth, the crust is but a thin sheath covering the planet, much like the outer skin of an apple or the bumps on the outer surface of a basketball. Earth's crust can be either oceanic or continental. The continental crust is less dense, thicker, and composed of lighter minerals than oceanic crust. Like the portion of an iceberg that is hidden beneath the sea surface, the continents also have a thick "root" of buoyant rock that keeps them sitting high above the underlying layer.

Beneath Earth's crust lies the *rocky mantle,* a thick layer of dense, sometimes semimolten rock, rich in iron and magnesium. A dividing line, called the Mohorovicic discontinuity, or simply the Moho, marks the division between the crust and mantle. Based on seismic wave data and rocks uplifted on land or collected at sea, it is believed that the mantle's composition is similar to that of a mineral called peridiotite (a light- to dark-green silicate (SiO_2) rock, rich in magnesium and iron).

Below the mantle, some 2900 kilometers from the surface, is the *core.* Calculations from measurements of gravity, earthquake data, and the composition of meteorites suggest that the core is composed of a very dense metallic material, probably a mixture of iron and nickel.

Increasing temperature and pressure alters the physical state of Earth's internal layering. Laboratory experiments suggest that the temperature of the outer core hovers around a blazingly hot 5000°C, nearly as hot as the sun's surface (Lamb and Sington, 1998). Outside of the core, Earth's layers insulate the surface from its hot interior, like the lining of a thermos. Much of what we know about the physical characteristics of Earth's interior comes from the study of how seismic waves or vibrations generated by earthquakes pass through the planet. Two important types of seismic waves are primary or P-waves and secondary or S-waves. Primary waves are compressional and pass through materials by jiggling molecules back and forth, parallel to the direction of travel. If one stretches a Slinky and then firmly taps one end, a wave travels down the spring moving individual sections to and fro; this is like the compressional motion of a P-wave. An important property of P-waves is that as the density of the surrounding material increases, so does their speed. P-waves can pass through both solids and liquids. Shear or secondary waves propagate by deforming a material or shifting the molecules from side to side. S-waves can pass through solids, but not liquids, because true liquids cannot deform. By studying how both P-waves and S-waves travel through Earth, scientists can estimate the relative hardness or fluidity of Earth's internal layers.

Near the surface, the crust and upper mantle together form a rigid, hard layer called the lithosphere (from the Greek word *lithos* meaning "stone"). In the analogy of a Milky Way bar, the lithosphere is the thin, brittle layer of chocolate on top. The lithosphere extends from the surface to a depth of approximately 100 km beneath the oceans and 100 to 200 km below the continents. Below the lithosphere, a relatively thin zone exists in which both P-waves and S-waves slow. This low-velocity layer, approximately 100 kilometers thick, is called the asthenosphere (from the Greek word *asthenes* meaning "weak"). The slowing of seismic waves in the asthenosphere suggests that it is partially molten or fluid-like, able to deform plastically, something similar to tar, asphalt, or as in the Milky Way model, caramel. Below the asthenosphere, the mantle appears to harden, but its exact nature remains uncertain. Seismic discontinuities, or changes in seismic wave speed, occur at depths of 410 and 670 kilometers within the mantle and are believed to reflect changes in mineral structure (not composition). New research indicates that at the base of the mantle, the core-mantle boundary, a 5- to 50-kilometer-thick layer exists in which seismic velocities are also reduced. Scientists now believe that a partial melt or fluid-like molten mass may exist at the core-mantle boundary—a surprising theory that suggests a more complex region than previously hypothesized.

Research also indicates that S-waves cannot pass through the outer part of Earth's metallic core; therefore, it is believed that the outer portion of the core is liquid and the inner, solid. Earth's magnetic field is thought to derive from the planet's rotation about its axis and the subsequent motions of the outer, metallic, liquid core. The transition from liquid to solid at the boundary of the outer-inner core probably results from increases in both temperature and pressure. Based on new seismic data, some scientists believe that the inner core is not a simple metallic solid as once speculated, but may be rotating independent of the planet; it may be softer and contain iron crystals that are unevenly distributed or oriented throughout the core (Monastersky, 1998). These new findings are controversial. Further research may or may not lead to confirmation, but undoubtedly as technology improves and scientists continue to study Earth, we will learn more about the interior and very center of our planet.

Angry Borders

Earth's surface is divided into about 15 lithospheric plates that are internally rigid and overlie the more mobile asthenosphere (Figure 31). The plates are irregular in shape, vary in size, and move relative to one another over the spherical surface. A single plate can contain oceanic crust, continental crust, or both. They are continually in motion, in relation to each other and to Earth's rotation. At their boundaries, the plates constantly jostle and grind against one another, creating huge mountain chains or deep-sea trenches. It is also here, at the plate's angry borders that Earth's fury is at its peak, and the majority of the world's earthquakes, volcanoes, and tsunamis are generated.

Divergent Boundaries. A divergent boundary occurs where two lithospheric plates are moving away from one another. The midocean ridge

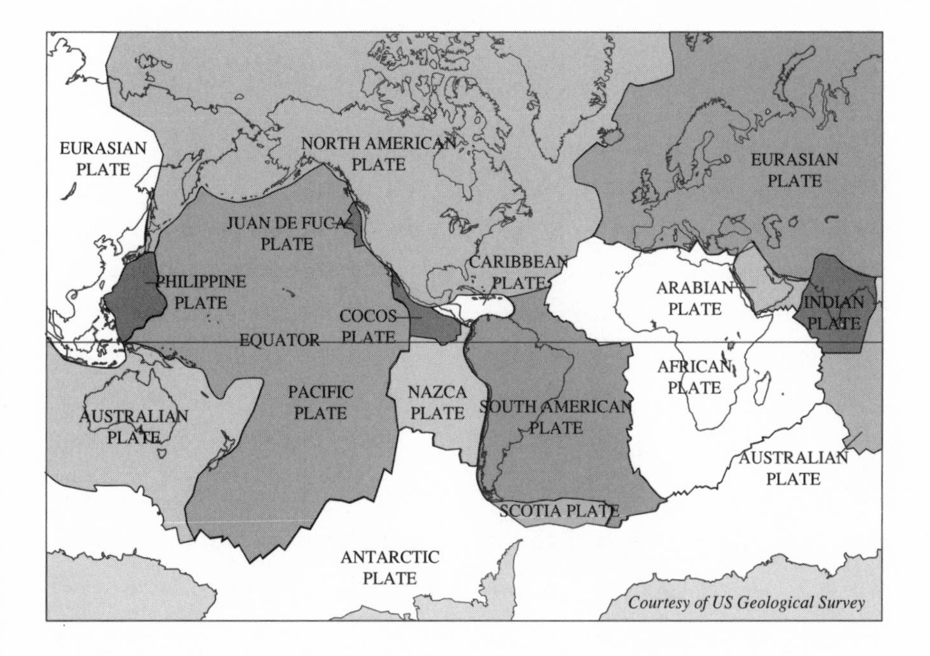

Figure 31 Earth's surface divided into its tectonic, or lithospheric, plates.

system, the most extensive mountain chain on Earth, is a consequence of plate divergence. At the crest of a midocean ridge, lithospheric plates move apart and molten rock from deep within the planet wells upward. When it is beneath Earth's surface, molten rock is called *magma;* when it erupts or oozes out above ground, it is called *lava.* Magma is generally a mixture of melted or crystallized minerals and dissolved gases. It is typically less dense than the surrounding materials, so buoyancy drives it upward. At a midocean ridge, magma rises toward the surface and erupts onto the seafloor to create new ocean crust. Here, deep in the sea, when hot lava meets cold seawater, it cools very quickly and creates dark, glassy pillow basalts (see Figure 2). The separation of plates at a midocean ridge and the creation of new oceanic crust are called seafloor spreading.

Seafloor spreading occurs intermittently and at varying rates. In the Pacific Ocean, along the East Pacific Rise, new seafloor is created at a rate of approximately 6 to 17 centimeters per year. In contrast, in the Atlantic, along the Mid-Atlantic Ridge, spreading is slower, an estimated 1 to 3 centimeters per year. Variations in heat flow, the chemical composition of upwelled magma, and the structure of a ridge along its axis appear related to the spreading rate. At the Mid-Atlantic Ridge, a slow-spreading ridge, the magma is blocky, relatively viscous, and forms a steep, rocky terrain with a topographic low or valley along the rift axis. At the East Pacific Rise, a fast-spreading ridge, molten material is thinner, less viscous, and forms a flat, broad ridge with a topographic high at its center. Scientists speculate that beneath fast-spreading ridges there exists a narrow zone of high heat and melting. Seismic evidence and three-dimensional imaging suggest that 1 to 2 kilometers beneath the East Pacific Rise lies a thin horizontal layer of molten material that feeds the spreading center. At slow-spreading ridges the axis appears to be cooler, thicker, and subject to greater faulting and earthquake activity.

The Mid-Atlantic Ridge runs smack through the middle of Iceland. Consequently, in Iceland scientists are availed an unparalleled look at the processes of rifting along a slow-spreading mid-ocean ridge. Studies indicate that rifting occurs by a slow widening and sinking at the ridge axis, until a breaking point is reached and fractures occur. Cracks begin to form parallel to the rift, earthquakes jolt the region, and lava erupts through some of the fissures (Decker and Decker, 1998). Rifting events may be tens of kilometers long and they tend to occur infrequently at any one spot. Along the world's mid-ocean ridges and their associated fracture zones are sites of active hydrothermal activity, known as deep-sea vents. The relatively recent discovery of these vents, in the late 1970s, has had a profound impact on our understanding of the oceans and the very origins of life. (Aspects of the geology of vent deposits are given later in the chapter and more detail on the biological nature of vent communities will be discussed in the following chapter.)

Convergent Boundaries. New crust continually forms at the midocean ridges, but Earth's size has not changed significantly for millions, if not, billions of years. So somewhere on the planet's surface, crust is being destroyed at the same rate as it is created. We now know that crust destruction occurs where two lithospheric or tectonic plates collide. There are essentially three types of collision.

1. *Continental-continental collisions.* When two plates collide and each is composed of continental crust, towering mountains are created. Like two crashing automobiles, the plates' edges are smashed and crumpled upward. When India crashed into Asia some 50 million years ago, the crumpling and crashing of the edges of the plates created the towering Himalayan Mountains.
2. *Oceanic-oceanic collisions.* The Marianas Trench off the coast of the Philippine Islands in the Pacific Ocean is some 11 kilometers deep, the deepest site in the sea. Beneath the Marianas Trench two plates of oceanic crust are colliding, the Pacific plate and the Philippine plate. When two oceanic plates converge, usually the older, denser plate is driven beneath the younger, less dense plate. (As ocean crust ages and spreads away from a midocean ridge, it cools and its density increases.) However, exceptions do occur; the younger Caribbean plate is inexplicably being driven beneath the older South American plate. When one plate is forced beneath the other, it is called subduction, and the area in which this occurs is called a subduction zone (Figure 32). Ocean trenches are the surface expression of a subduction zone. During the subduction process, water deep within Earth is thought to be an important lubricating agent, allowing one plate to slide over another. Even so, the subduction of

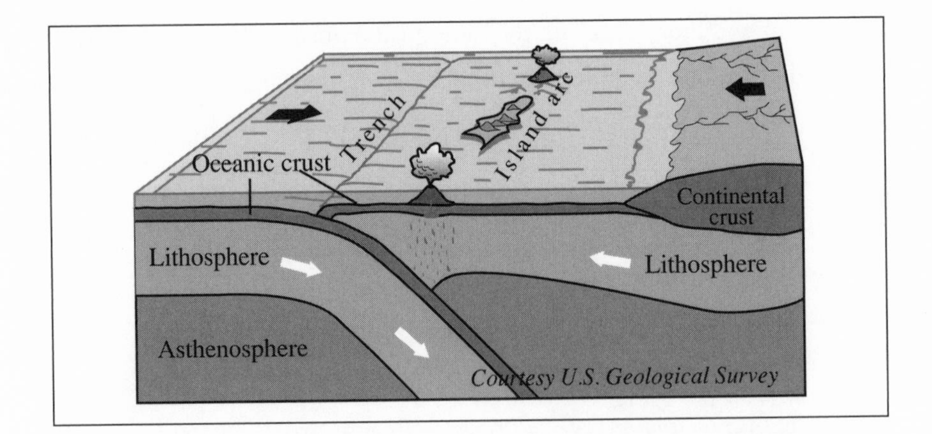

Figure 32 The collision of two tectonic plates, each with oceanic crust on its leading edge.

Earth's crust produces the planet's largest and most devastating earthquakes. These earthquakes and the associated deformation of the seafloor can also spawn towering tsunamis. Additionally, high temperature deep in a subduction zone melts the downgoing slab and generates molten rock. Driven by buoyancy, hot magma flows upward through fractures, in the overlying rock, and can erupt at the surface to form a chain or arc of active volcanoes behind the subduction zone. An arc of volcanoes known as the ring of fire rims the Pacific Ocean (Figure 33). Seventy-five percent of Earth's active volcanoes and most of the planet's earthquakes and tsunamis occur within the Pacific's infamous Ring of Fire.

3. *Oceanic-continental collisions.* Since oceanic crust is denser than continental crust, when the two collide oceanic crust is forced downward beneath continental crust. For instance, at the Peru-Chile trench the oceanic Nazca plate is being driven beneath the South American continent, part of the South American plate. Behind the subduction zone, great upheavals of the land and slow

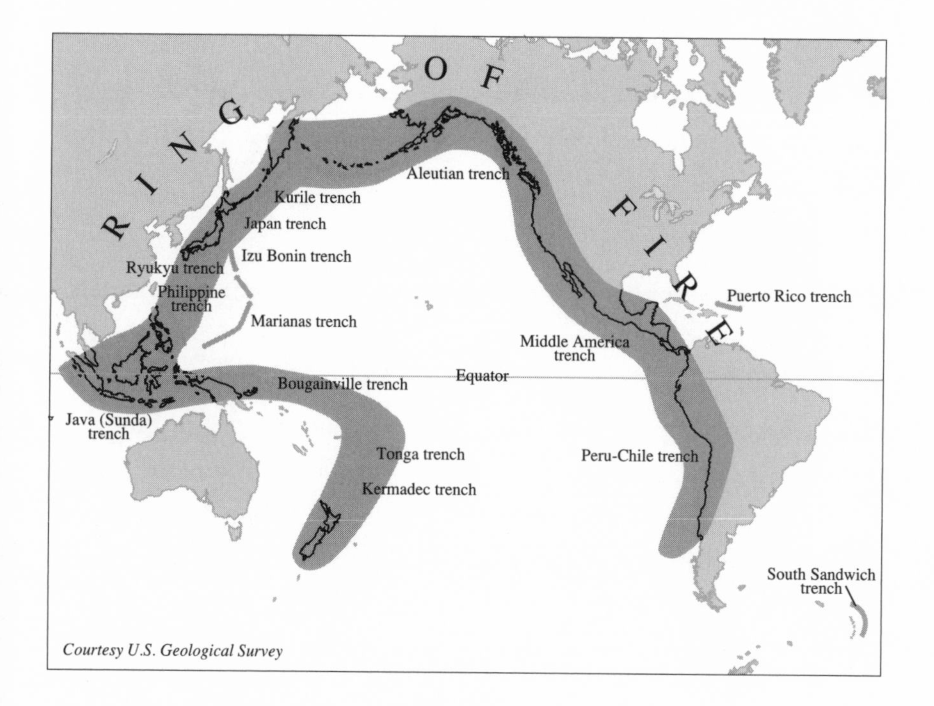

Figure 33 The infamous Pacific Ring of Fire.

continuous uplift have created the lofty Andes Mountains. During collisions of oceanic and continental crust, or oceanic and oceanic crust, some of the sediment and rock on the downgoing slab may be scraped off and pasted onto the overriding plate. The island of Barbados is built on a wedge of material scraped off the Caribbean plate as it dives beneath the South American plate.

Transform Faults. Transform faults are where two plates slide in opposite directions past one another. Across the midocean ridges transform faults create numerous fracture zones. Shallow earthquakes are common along transform faults. The most famous—or infamous, as the case may be—transform fault is California's San Andreas fault. Here, the Pacific plate, which includes part of California, is moving approximately 1 to 6 centimeters per year northwest against the southeasterly moving North American plate, which includes the rest of the state. If plate motion continues, sometime in the distant future San Francisco and Los Angeles will reside at the same latitude.

Hot Spots

Much like a conveyor belt, the tectonic plates moves over Earth's surface. At some point in its history a plate may pass over a stationary thermal plume, or hot spot. For some reason, there are fixed places inside Earth's mantle that are unusually hot. Here, rising heat and erupting magma generate a series of volcanic features such as seamounts or volcanic islands that trace the movement of the plate over the hot spot. The Hawaiian Island chain is the most well known product of a hot spot (Figure 34). As

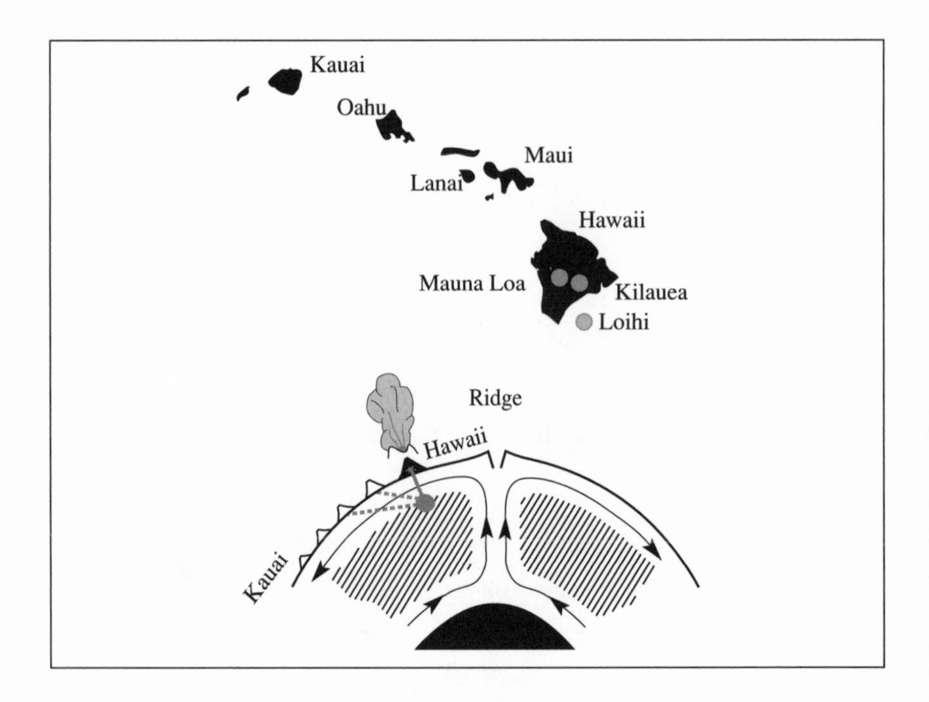

Figure 34 Hawaiian Islands and an underlying hot spot.

the Pacific plate moves over an underlying hot spot, the Hawaiian Islands are created. Hawaii is a relatively recent hot-spot creation, but now a new submerged volcano, named Loihi, is forming to its southeast. By dating rocks on the islands, scientists have determined that the Pacific plate has moved an average of 8.6 centimeters per year for at least 70 million years. A bend in the island chain suggests that some 40 million years ago, the movement of the Pacific plate changed direction, from north to northwest.

Scientists now believe that hot spots may be the dominant means by which oceanic crust is modified between creation at midocean ridges and subduction in deep-sea trenches. In addition, because the lava at hot-spot volcanoes is chemically different from that at midocean ridges, hot spots are thought to originate from thermal plumes deep in the mantle, below the asthenosphere and maybe even at the core-mantle boundary. No obvious pattern in the distribution of presently active hot spots has been determined though some concentration seems to occur in regions away from subduction zones. Hot spots occur less commonly under the continents. The famous geysers, boiling mud pools, and steaming landscapes of Yellowstone National Park are thought to result from a hot spot underlying the North American continent. Research also suggests that some 100 million years ago, hot-spot activity was five to ten times greater than it is today and may have had a significant impact on ancient ocean temperatures, sea level, and climate. Many puzzling questions about hot spots remain unanswered. Scientists are actively investigating the source of hot-spot heat and magma, the causes for change in hot-spot activity, the controls on their spatial distribution, and their link to climate change.

The Force

Why and how lithospheric plates move has confounded scientists for decades; it was one of the most troubling aspects of the early continental drift and plate tectonic theories. Even today, our understanding of the forces that drive plate motion is rudimentary. Much of the difficulty arises because we have no method to directly view the planet's interior or test relevant theories. None of the presently proposed mechanisms of plate motion seem to explain all aspects of plate movement, but for now, they are our best guess based on the available evidence.

Plate motion appears to be driven mainly by convection within Earth's mantle layer and pull from plate subduction (Figure 35). As discussed earlier, the asthenosphere, a thin layer in the upper mantle, is believed to be partially molten. Heat from deep within the planet is thought to cause very slow convection currents (remember the Lava Lamp) in the partially molten asthenosphere (or maybe deeper). The heat source for convection within the asthenosphere comes from deep in Earth's interior, fueled by the decay of naturally radioactive materials (e.g., uranium, plutonium, thorium) and heat from the early formation of the planet. In a simplified conceptual model of what must be a spatially

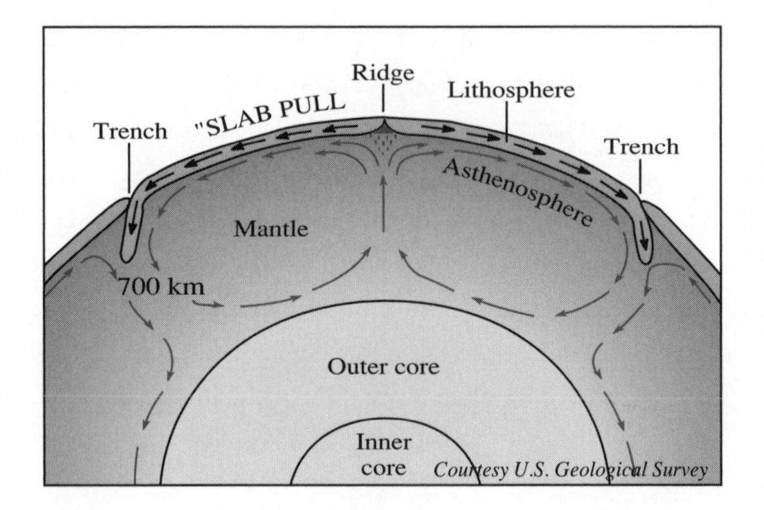

Figure 35 Convection and slab pull in the Earth's mantle and asthenosphere.

complex and dynamic system, we envision a series of convection cells within the asthenosphere. Uneven heating causes thermal plumes to rise at midocean ridges and cooling near the surface creates descending plumes at subduction zones. In between, the asthenosphere moves horizontally from beneath a spreading center—a ridge—to a subduction zone—a trench. Friction between the lithosphere and the asthenosphere acts like glue, and the lithospheric plates are dragged along by the motion of the underlying asthenosphere. It is difficult to conceptualize how Earth's rocky mantle could flow almost like a fluid. Imagine the children's toy, Silly Putty. When molded tightly into a ball, a large mass of the squishy stuff will bounce like a solid off a hard surface. However, if placed in one spot for an hour or more, the silly putty, like a very thick fluid, will gradually spread out over the surface. It acts like a solid over the short-term and a fluid over longer time periods. Mantle convection may be similar, flowing very slowly over a very long period of time.

At subduction zones, gravity pulls the slabs of cold, dense oceanic crust down into the mantle. This process is now believed to play a major role in plate motion by pulling the entire plate along with the descending slab. In the past, scientists also speculated that the formation of new crust at the spreading ridges pushed the plates apart. Now most believe that very little horizontal force is actually created during spreading and that it is minor when compared to the "slab pull" at the subduction zones or convection within the asthenosphere. The source of plate motion will undoubtedly be the focus of scientific study and debate well into the future, and there remains much to learn about Earth's changing surface and internal structure. For instance, what controls the rate of seafloor spreading or the direction of plate movement? Why do changes in motion or initial rifting occur? Are there really convection currents in the asthenosphere, or are there other forces that help to drive Earth's geologic motions?

Seafloor Topography

In another Jules Verne classic, *20,000 Leagues under the Sea,* the author again transports the reader to a place far from our terrestrial world, the deep sea. As with our understanding of Earth's interior, our knowledge of the seafloor is severely limited by our inability to simply observe its nature. Looking through the air, for the most part transparent, we can easily view the thick forests, steep canyons, or towering mountains that characterize Earth's landscape. With space-age technology, we can now even view the barren, rock-strewn surface of Mars. But the ocean is an opaque medium, and the nature of the seafloor remains mysterious, hidden beneath a dark and imposing blanket of water. Deep-sea vents and their thriving communities were completely unknown till some 20 years ago, and now we find them throughout the world's oceans. Only a select and lucky few have ever observed first-hand the floor of the deep sea.

Breakthroughs in the viewing and mapping of the seafloor have come with advances in marine technology. The first revelation came with the advent of echo sounding as a means of measuring depth in the sea and mapping its distribution. Ships began traversing the sea with scientists aboard measuring depth continuously along their track. In the early 1950s, under the tutelage of Maurice Ewing of the Lamont Geological Observatory at Columbia University, graduate student Bruce Heezen and a young geologist, Marie Tharp, began to compile all of the available sounding profiles across the Atlantic. Their herculean efforts lead to the first set of comprehensive and detailed maps of the seafloor. Heezen and Tharp's depictions are remarkably accurate and are probably the most widely distributed maps of the ocean's bottom. In 1996, Walter Smith and David Sandwell of Scripps Institution of Oceanography published a spectacular new map of the seafloor (Color

Plates 5 and 6). This updated blueprint of the ocean's undersea architecture contains unprecedented detail and has revealed seafloor structures in areas originally thought featureless. For instance, the map documents a previously unknown 1600-kilometer chain of seamounts in the South Pacific. The new image by Smith and Sandwell is based on all available depth soundings as well as gravity data obtained by orbiting satellites (Smith and Sandwell, 1997).

On any map of the seafloor, the submerged but mountainous mid-ocean ridges and their offsetting fractures jump out at the observer. Another feature that stands out are the deep, narrow scars around the edges of some ocean basins. These are the deep-sea trenches marking zones of subduction. They are V-shaped with relatively narrow, steep walls and are typically some 100 kilometers wide and hundreds to thousands of kilometers long. The deepest trenches occur in the Pacific: the Marianas, 10.9 km; the Tonga, 10.8 km, the Philippine, 10 km. Trenches are shallower where sediments spill into and pile up within the undersea crevasses: the Puerto Rico Trench, 8.6 km deep.

Between the continents, trenches, and mid-ocean ridges lie broad, flat undersea plains speckled with underwater peaks and seamounts. This is the realm of the abyssal plain, the flattest region on Earth. Here, sediments raining down from above bury the rough, underlying volcanic terrain and form a smooth, low seafloor that averages about 3 to 5 kilometers in depth. In some areas, the abyssal plains are dotted with domes or elongated hills made up of volcanic rock with a thin veil of overlying sediment. Seamounts, which were once active volcanoes, may rise steeply above the seafloor and occur singly, as a chain, or in a cluster of peaks. Some seamounts are flat-topped, having once risen above the sea where erosion by waves removed their jagged peaks or coral reefs covered their surfaces. As techniques to observe and sample the deep seafloor improve, we will gain a more detailed picture of the seabed and smaller features that were unseen before will undoubtedly emerge.

Along the edge of the ocean lies the continental margin, the interface between land and sea. Here the land begins to slope into the abyss, sediment flows from the continents offshore, and ancient rivers and underwater avalanches carve out deep submarine canyons. In some areas, the land slopes gradually into the sea, forming a broad, flat shelf, while in other settings the transition is quick and narrow. The continental shelf, a flat brim bordering the ocean, averages about 60 kilometers in width, though it can be as wide as 1000 kilometers in the Arctic Ocean or as narrow as a few kilometers along the Pacific coast of North and South America. At a depth of about 130 to 200 meters, the continental shelf steepens to form the continental slope. Though it is called "the slope," the steepness of this slope is often so small that if you were standing on it and looking seaward, it would in fact appear perfectly level. Over the ages, sediments worn from the land pile up beneath the continental shelf and slope, and in some areas, huge submarine canyons cut deep scars into their surface. These undersea canyons act as chutes, transporting sediments from the land into the sea.

The formation of submarine canyons has been the subject of great geologic debate over the years. Scientists now believe that submarine canyons are formed by river erosion during times of lowered sea level, by massive underwater avalanches of sediment and water, and by the slumping of canyon walls. At the base of the continental slope is the continental rise. Here the slope becomes less steep as sediments from the shelf and slope build up. The continental rise can extend into the deep ocean for hundreds of kilometers, reaching depths of some 4000 meters and the abyssal plains. The sequence of continental shelf, slope, and rise does not occur on all margins and may be altered by the presence of seafloor features, such as a deep-sea trench. The shape of a continental margin is influenced by its rock type, the past sea-level and tectonic history of the region, and the forces of wind, ice, and water.

Ocean Sediment and Rock

Ocean sediments cover most of the seafloor, forming a geologic cloak that hides the dark underlying volcanic crust. Undersea mountain peaks appear as if snow-topped, while the ocean's edges are often lined with sparkling grains of sand. The sediments of the sea reflect conditions that have prevailed for long, relatively quiet periods of time, or they may represent instantaneous and catastrophic events, such as a massive flood, powerful storm, or violent volcanic eruption. In either case, the seafloor unquestionably harbors a rich, though relatively untapped, record of ocean history and the planet's past. Marine geologists and oceanographers strive to unlock the wealth of knowledge about Earth that lies hidden within the ocean's sediments and rock.

Marine sediments are particles of organic or inorganic matter that accumulate in the ocean in a loose, unconsolidated form. Depending on their size, sediments are called mud (0.001–0.032 mm), sand (0.063–2 mm), or gravel (2–10 cm). Mud can be further divided into clay (0.001–0.004 mm) and silt particles (0.004–0.063 mm). The size, shape, and density of a grain determines how it moves in the ocean; rolling and jumping on the seafloor or carried suspended within the water's flow. Grains can be round like a small bead, flat like a flake of oatmeal, or angular at their sides. Some sediment grains take on a shine as if polished, while others remain dull over time. The nature of the grains often provides a clue to whence the sediment came. Ocean sediments can also be well-sorted, made up of grains all nearly the same size, or they may be poorly sorted, containing a great mix of sizes. Over time compaction, crystallization, and cementation can transform the sea's loose sediments into hardened rock. Much of the ocean's sediment originates within the sea itself, while the rest is carried in on the backs of wind, water, and ice.

Agents of Transport

Winds high over the land can carry dust-sized particles over great distances toward the sea. In the Caribbean, hazy days often signify the airborne onslaught of fine, reddish dust blown in from the Sahara Desert. In the Bahamas some of the rocks are red, colored by iron carried in by dusty winds coming from the deserts of Africa. The action of wind on a sediment grain often lends it an angular shape and frosted appearance, much like the sandblasting of glass. Winds can also spread the ash from an erupting volcano around the globe and blanket the oceans with a layer of fine volcanic debris. Layers of volcanic ash found in deep-sea sediment cores are used as markers for dating and locating ancient volcanic eruptions.

Throughout the ages, rivers have carried vast quantities of sediment into the sea. Upon entering the ocean, rivers tend to drop their coarse load near the shore, while currents carry finer materials farther out to sea. At their mouths, debris from the land collects and can form a delta. Currents, waves, and tides may spread a river's wealth out over the continental shelf, slope, and rise. Where waves roll and tumble the sediment, grains may take on a rounded, often polished appearance. The famous big-wave surfing beaches of Hawaii are lined with sands of shiny, rounded grains (Figure 36). During times of low sea level, rivers erode and transport sediments far offshore. However, when the sea is relatively high, like today, estuaries fill with debris from the land.

In polar regions, the bulldozing action of mighty glaciers pushes ice and rock toward the sea. Upon meeting the edge of the ocean, ice and snow melt or calve off in a spectacular collapse, thereby dumping the glacier's sediments into the ocean. The powerful erosive force of a glacier can also carve out deeply incised valleys or fjords or pile sediment into a

Figure 36 Rounded sand grains from a well-known surfing beach in Hawaii. (E. Prager.)

huge ridge of jumbled, poorly sorted boulders, gravel, and sand. The curling arm of land off the east coast of North America, known as Cape Cod, is a pile of rocks and sand built by an ancient bulldozer of ice. Additionally, melting icebergs may dump far from land the sediments frozen within their icy grip.

Sediments also enter the deep sea through the eternal influences of gravity. Gravity pulls sediments and rocks on the sloping seafloor downward into the dark abyss by a slow creep or a rapid giving away of the surface. Oversteepening of a slope, an earthquake, or the release of trapped gas (a seafloor burp) can trigger large blocks or huge areas of sediment to instantaneously give way and slump downward. Slumping of seafloor sediments poses a particular hazard to offshore oil rigs in the Gulf of Mexico where huge quantities of sediment coming from the Mississippi River have piled up on the seafloor and periodically give way. Turbidity currents are equally dangerous to offshore operations. These undersea mudslides are triggered the same way as sediment slumps, but are fast, short-lived flows of a sediment-water slurry.

Turbidity currents gained recognition in 1929 when an earthquake shook the seafloor south of Newfoundland and triggered a massive undersea avalanche. The avalanche or turbidity current wreaked havoc on transatlantic telegraph cables lying submerged off the Grand Banks and revealed much about these powerful undersea events. Eight cables snapped almost simultaneously near the earthquake's origin, and some 13 hours later they continued to break, the farthest 675 kilometers to the south. Based on the timing of when the cable connections were broken and communications quit, and their spacing on the seafloor, geologists estimated that the turbidity current raced seaward at some 40 to 55 kilometers per hour. It left behind a 1-meter thick deposit of sediment, covering over 100,000 square kilometers (Heezen and Ewing, 1952; Pinet, 1992). The debris laid down by a turbidity current tends to form a distinct series of sediment layers. When identified in present-day sediments or ancient rock formations, these distinctive layers indicate a once-racing avalanche of mud and water.

Sediments of Life

Much of the ocean's sediment comes from within the sea itself. These are the shells, skeletons, and teeth of marine organisms whose life in the ocean has ended but whose hard parts remain preserved as sediment. Deep-sea sediment of biological origin contains mostly small shells of calcium carbonate or silica. Because of their fine texture and squishy nature, the soft sediments of the deep sea are called oozes and are specifically named for the organisms that make up their bulk, for example, foraminifera ooze, radiolarian ooze, etc. The most prevalent ooze contains an abundance of small, spherical calcium carbonate globes; the shells of foraminifera (see Figure 14). Coccolithophores and pteropods also add their delicate shells to the calcium carbonate sediments of the seabed. Deep-sea ooze of silica is created by the accumulation of vast numbers of small siliceous shells, such as those produced by diatoms, radiolarians, and a tiny planktonic creature called a silicoflagellate. In shallow-water regions, fragments of shell and skeleton from coral, shellfish, echinoderms, crustaceans, algae, and fish add to the biological debris flooring the sea.

Another important biologic component of the ocean's sediment is organic matter, literally the dead or dying bodies of marine plants and animals. Most of the time, after scavengers and decay, little of this material actually makes it to the seabed. But during times of prosperity in the sea, when the production of marine organisms is high and rapid, great amounts of organic material can be incorporated into the sediments. A dark-brown, gray, or black color is generally imparted on sediments rich in organic matter. Over time and with burial deep in Earth, the organic remains of tiny sea creatures can create some of the most valuable and politically powerful materials on the planet: oil and gas.

Land Contributions

Sediments derived from the land are also quite abundant in the ocean, particularly close to shore. Most of the sea's land-derived, or terrigenous, sediments are associated with the inflow of large rivers. However, fine bits of land-derived clay are one of the few materials found in the ocean's greatest depths. Although the same flakey grains of brown-red clay occur throughout all of the world's oceans, in most places other sediment types overwhelm the tiny pieces. There are four major types of clay in the sea, and each holds a clue to its origin. Two of the clays in the ocean, illite and chlorite, are typically found where rivers run through high mountainous regions. Chlorite is most commonly produced by physical or mechanical weathering, such as the freezing and thawing processes prevalent in cold to temperate regions. Illite most often occurs close to a river's mouth. The other two clays of the sea, kaolinite and montmorillinite, are products of alteration. Kaolinite forms as a result of extensive chemical weathering characteristic of hot, wet tropical environments. And montmorillinite forms when volcanic sediments are chemically altered. Together, the land and biologically derived materials compose 95 to 99 percent of all marine sediments.

Along with the clays and biologic oozes of the deep seafloor, the ocean also contains volcanic ash, fragments of meteorites, and a variety of other geological creations that are produced by chemical and biological reactions within the sea. Most sediment is concentrated near the continental margins; very little actually makes it into the deep sea. So in the ocean's deepest reaches, things that enter infrequently or at a very slow rate can be found. During the *Challenger* expedition, deep-sea sediment samples reportedly contained small, black magnetic particles. These glassy, teardrop-shaped sediments were bits and pieces of meteorite debris that fell into the sea or came from impacts on land. Such cosmic sediments in the ocean are small in total volume and are estimated to accumulate very, very slowly, only 2 millimeters per 100 million years.

Chemical and Biological Creations

A wide variety of other ocean sediments, deep and shallow, are created by chemically induced and biologically mediated precipitation from seawater. At sites of deep-sea hydrothermal activity, hot mineral-rich water flowing out of fractures in the seafloor mixes with the ocean's frigid bottom water. Instant chilling of the hot plume water causes metal oxides and sulfide materials to rapidly precipitate. At hot chimneys (270–400°C), mineral precipitation makes the water look black, lending them the name black smokers (Figure 37). At cooler sites (25–270°C), water gushing

Figure 37 Black smoker at a deep-sea vent. (Courtesy of OAR/National Undersea Research Program, NOAA.)

from the vents appears white in color; these have been dubbed white smokers. Minerals precipitating out of the "smoking" vent as water rapidly accumulates on the surrounding seafloor, often creating dark, massive, metal-rich chimney structures around the vents. Chimney structures can grow amazingly fast, some 30 centimeters per day (Humphris and McCollom, 1998) and reach towering heights. On the Juan de Fuca ridge, off the west coast of Seattle, one chimney has grown to over 15 stories in height and was dubbed Godzilla by the scientists who discovered it (Broad, 1997). Precipitation of mineral-rich materials can clog up a vent, like a stopped-up drain, and reduce or stop plume emanations. In 1998, using a remotely operated vehicle, scientists were able to collect large chimney structures from deep beneath the sea for the first time. Intense research on the recovered samples is now underway to learn more about vent processes, their mineral accumulations, and the associated biologic communities.

Deep-sea vent environments are thought to have originated early in the ocean's evolution, so with a modern analog in hand, scientists are now scouring the continents in search of fossil, metal-rich vent deposits. So far, ancient vents and their associated sulfide ores have been found in the raised rocks of places such as Oman, Cyprus, and Newfoundland.

Evaporites and an Undersea Lake of Brine. Evaporites are rocks or sediments formed by evaporation. The same process that leaves a layer of salt on one's skin after an ocean swim can create massive deposits of evaporite minerals such as salt, gypsum, or anhydrite. In the past, particularly during the late Mesozoic era some 150 to 65 million years ago, thick evaporite deposits were created when broad, shallow seas periodically dried up. Beneath the present Gulf of Mexico, buried some 3700 meters (12,000 feet) below the seabed, lie enormous deposits of salt created by ancient episodes of evaporation. The weight of overlying sediments has compressed and molded the salt into huge domes. Above the salt domes, in the deep water of the Gulf of Mexico, natural gas, oil, and salt-rich fluids or brines seep out of cracks in the seafloor. In the 1980s, scientists went to this region to investigate the effects of natural seeps on bottom-dwelling creatures. They expected to collect fish and other species sick from the effects of oil and brine. To their great surprise sampling devices returned overflowing with healthy, thriving tube worms similar to those found at deep-sea vents (McDonald and Fisher, 1996). Submersible dives 128 kilometers off the coast of Louisiana revealed the cause of this amazing abundance of life and an extraordinary deep-sea find. Similar to the communities associated with hot deep-sea vents, a luxuriant assemblage of tube worms, clams, mussels, and other organisms were living clustered about cold seeps of gas and oil (Color Plate 1). In addition, a strange pool of thick brine, some four times as salty as seawater, was found in a crater created by the release of buried gas. The edges of the undersea brine lake were lined with organisms living off bacteria thriving on seeps of methane and hydrogen sulfide. Looking through the submersible's portholes, researchers saw that organisms such as fish or crabs that came to feed along the fertile shores of the brine pool sometimes fell into its toxic sludge and were instantly pickled, making it a sort of underwater La Brea tar (salt) pit.

The domes beneath Louisiana are heavily mined and provide an important source of salt for everyday use in the modern world. Salt is also mined in the tropics where island residents capitalize on the flow of the tides and power of the sun to harvest the sea's crystal treasure. Seawater, driven by the tide, is funneled into and trapped within shallow, broad holding pens. Under the tropical sun and heat, the water evaporates and a crystal blanket of shiny, white salt is left behind and harvested.

Manganese Nodules. Deep within the Pacific Ocean, fields of dark, potato-shaped rocks blanket the seafloor. These manganese nodules grow round as manganese, copper, nickel, cobalt, and iron precipitate out of seawater and coat their surface. Any piece or particle on the seafloor can serve as a nucleus for a growing nodule, and in some instances, even an ancient shark's tooth has been found at a nodule's core. Manganese nodules grow slowly, only 1 to 4 millimeters per year; many are just 1 to 10 centimeters in diameter. Because their growth is slow, they only form in areas where the rate of sediment accumulation is also slow; otherwise they would soon be buried. Manganese nodules are most abundant in the deep Pacific, but their distribution is quite patchy. In some areas the dark undersea potatoes lie scattered about, while in other places they cover the seafloor in a black bumpy carpet. Some of the patchiness may be due to periodic burial by sediments.

For years, the origin of the manganese and other metals in the nodules was a mystery; seawater typically contains little of these elements. It is now believed that much of the dark, metalliferous material originates in the metal-rich plumes of superheated water at deep-sea vents. However, the exact process of how metals precipitate out of seawater to form a rounded nodule deep in the sea is uncertain. Currents strong enough to roll the nodules occur infrequently, if at all. But because they grow very slowly, rolling only once in a while might be sufficient to produce a rounded shape. Some scientists believe that bacteria promote mineral precipitation or that other organisms crawling and burrowing on the seafloor may unknowingly roll the nodules. This is just one more of the sea's deep, dark, and in this case, rounded secrets. So far, no economical means of mining manganese nodules has been found, but their potential value has spurred great international debate over territorial rights in the sea.

Fostering Phosphates. After the *Challenger* expedition was completed, scientist John Murray became a wealthy man and the British Crown's Treasury recovered the costs of the voyage. These riches came not from the wealth of knowledge unlocked by the voyage, but from a deposit of phosphate discovered on Christmas Island during the expedition that was later mined. Within the ocean, phosphates typically accumulate on the seafloor beneath regions of upwelling or rich growth. They can precipitate from seawater chemically, with the help of microorganisms or as an alteration product of organic matter or fine-grained carbonate rocks. Phosphate-rich undersea sediments commonly form black or brown rounded grains or irregularly shaped crusts. Fossilized shark teeth sometimes found among the grains of sand on a beach obtain their black color from the recrystallization of the teeth's calcium phosphate. Fossil phosphate deposits, such as those that occur in Florida and Georgia, are heavily mined for use in fertilizers. Seabird excrement, called *guano,* is also extremely phosphate-rich (and stinky). Areas that host large seabird colonies sometimes accumulate thick deposits of valuable phosphate-rich guano. On Navassa, a remote island off the coast of Haiti, a supposed guano deposit was historically the cause of great territorial dispute and even a few murders. Claims to the island and its phosphate-rich material were based on the Guano Act, which pertained to claims to an island that was rich in guano. In 1998, a scientific expedition went to Navassa and geologists discovered that the "guano" was not guano after all, but a deposit of granular red-brown beads that appears to have originally formed underwater.

Blocks of Beachrock. Along the edges of the ocean, another example of chemical precipitation is easily observed and often walked upon; this is beachrock. Beachrock forms along the shore when minerals that are normally dissolved in seawater precipitate out and cement loose beach sand into hardened blocky rocks. Beachrock can only form where sediments remain relatively still over time, have a high permeability, and where seawater contains a lot of dissolved calcium carbonate or silica. Because beachrock forms relatively quickly, geologically speaking, cans, bottles, and other objects of the modern world can often be found cemented within the hardened blocks of rock along the shore. Off the Bahamas, an ancient deposit of beachrock now lies submerged beneath the sea, and its square blocks appear mysteriously road-like. Until geologists identified the stones as beachrock, some believed them to be the ancient roads of the lost city of Atlantis. Geologists identify and study beachrock more closely by creating a thin-section, a very thin slice of the rock glued onto a glass slide, and examining the rock's grains and cement under a microscope (Color Plate 11).

Oogling Ooids. In just two regions of the modern shallow sea, seemingly unnatural, round, shiny white beads of calcium carbonate are being created. These sediment beads are called ooids. They were common in ancient marine environments but now form only in the Bahamas and Persian Gulf. One of the main geologic formations underlying Miami and part of the lower Florida Keys is called the Miami oolite (ooid rock). Ooids form in shallow, agitated, warm waters that are supersaturated with calcium carbonate. When currents or waves pick up small particles or grains from the seafloor, they become coated with tiny needles of aragonite, a type of calcium carbonate (Color Plate 11). Over time, the coated grain is repeatedly picked up by the flow of water and a series of spherical, crystal layers form. Ooid formation may be facilitated by bacteria, with the microbes helping to precipitate calcium carbonate from seawater. On the other hand, the process of ooid formation may be strictly chemical, stemming from the oversaturation of the surrounding waters. Because movement by waves or currents is a requisite part of ooid formation, their size is generally limited to small, beadlike particles. In Florida, is desalination plants, where freshwater is created from saltwater, artificial ooids are sometimes created to remove minerals from the water. Ooids in the Bahamas form wonderful beaded beaches and create spectacular white sand waves that move with the flow of time.

Coral Reefs. Reefs are typically thought of as an environment dominated by biology, harboring one of the most diverse, abundant, and beautiful communities in the sea. In reality, coral reefs are but a thin veneer of living tissue. Lying below and responsible for the complex underwater topography that gives life and shelter to so many of the sea's creatures is a hardened, massive, and growing framework of rock. Beneath its living surface, a reef may be only 1 or 2 meters thick, or as in many regions, there may be a vast accretion of limestone, hundreds to thousands of meters deep. Given the right marine setting and appropriate ocean conditions, many corals can construct a massive skeleton of calcium carbonate. Calcium carbonate sediments produced from other organisms that live in and on a reef, such as the green calcareous algae Halimeda, fill the nooks and crannies within the coral framework. And coralline algae, encrusting "plants," provide a dense cement of calcium carbonate that tightly binds the reef together. Scientists use cores drilled through coral reefs to study their growth, history, and environmental changes that occur over time. In addition, cores drilled from an individual coral head can be sliced and x-rayed to reveal annual density banding that, like tree rings, can be counted to reveal age and changes in growth (Figure 38). Scientists have also found that when exposed to a "black light" (ultraviolet light), striking layers of white will fluoresce within a coral's skeleton. These bright bands are now believed to reflect times of high runoff, presumably caused by major storms, which bring large quantities of organic debris from the land into the sea, which gets incorporated into the coral's skeleton. Coral cores from the Florida Keys show that for decades prior to 1965 hurricanes hit the region every few years, but since that time very few have directly impacted the area.

Sediment Accumulation

In the shallow sea and at its edges, sediments can accumulate relatively fast, on the order of 5 to 30 centimeters per 1000 years, and reefs can grow even faster, up to 10 meters per 1000 years. In the deep sea, however,

Figure 38 (a) *Marine geologists drilling a coral head.* (b) *An x-radiograph showing density banding in a coral skeleton on left and fluorescing bands on the right.* *(Courtesy of Robert Halley, U.S. Geological Survey.)*

where sediments rain down in an endless underwater snowfall, accumulation rates are very slow, on the order of 1 to 25 millimeters per 1000 years. In fact, most of the material produced within the upper reaches of the open ocean never even reaches the bottom. Because the ocean's snow, tiny particles of shell or clay, falls very slowly, it may take 50 years for an individual particle to descend from the surface to the seafloor. On the way down, particles, shells and other debris may be eaten by organisms living within the water. However, being slurped up by a scavenging creature can actually promote faster sinking. These organisms remove the nutritive material from particles and excrete the rest as waste, in relatively larger, consolidated pellets. Whereas it may take tens of years for individual particles to fall to the seafloor, waste aggregates or fecal pellets sink much faster and can reach the bottom within just days or weeks. Getting incorporated into a pellet increases a particle's sinking rate and provides it with a protective coating of organic goo. The ocean's fecal pellets actually play an important role in getting sediments and organic material from the surface to the seafloor and in recycling nutrients.

Another factor that prevents most sediment, particularly calcium carbonate and silica shells, from reaching the seafloor is dissolution. Due to the chemistry of the oceans, silica tends to dissolve near the surface and calcium carbonate in the deeper sea. Also, because the shells of some planktonic creatures, such as the diatoms and pteropods, are delicate, they are easily crushed and rarely become preserved. For a biologic ooze to accumulate there must be a great abundance of organisms growing in the overlying water, and the depth and chemistry of the sea must be conducive to preservation.

Sediment Distribution

The mapping of ocean sediment began as early as the *Challenger* expedition and has continued ever since. Scientists today use a combination of low-tech and high-tech methods to retrieve, study, and map the sea's sediment. Just as in the early days of oceanographic research, sediment sampling is still often done with a towed dredge or a mechanical scoop dropped from a ship. Sediments may also be collected using SCUBA gear, submersibles, remotely operated vehicles, or a sediment trap. Sediment traps typically consist of an open funnel-shaped top attached to an underlying collecting cup. These simple but effective devices are placed on the seafloor or hanging within the water and left over time to collect marine sediments as they rain down from above. Marine geologists also use a variety of coring devices to penetrate the sediments and retrieve subsurface samples. Box cores are used to collect a wide, shallow, rectangular chunk of the seabed. These are especially useful for studying fauna that live within the sediments or delicate layering that may be disturbed at the edges of a narrow, cylindrical core. In shallow water, core samplers may be driven into the sediments by hand, with weights, using a piston, or with a mechanism to vibrate the core as it penetrates downward. In the deep sea, coring is usually gravity-driven in soft sediments (pistons and weights may aid deeper penetration) and rotary, to drill through rock. The first major seafloor coring was done by the Deep Sea Drilling Project (DSDP) and is now being accomplished by its successor, the Ocean Drilling Program (ODP). Today, the ODP has drilled throughout the world's oceans, including in water depths of almost 6000 meters in the oldest part of the Pacific Ocean, and cores have reached some 2111 meters below the surface of the seabed (Color Plates 4, 5, and 6).

Locating drill stations, ship positioning, and relocating earlier core sites has been greatly advanced by the creation and use of the global positioning system (GPS). Originally developed by the military to precisely locate positions on Earth, GPS has truly revolutionized almost every avenue of scientific study in the sea. GPS lets scientists accurately map and relocate almost any sampling site in the oceans. Global positioning basically works by measuring the distance and angle from any location on

Earth's surface relative to several orbiting satellites. Using one receiver on Earth's surface, the precision of GPS is on the order of meters; with several receiving stations, it is accurate to within an amazingly precise, few centimeters. However, GPS does not work underwater, so positions must be located at the sea surface and then correlated to sites on the seabed.

With these tools in hand, researchers have scraped, probed, dredged, drilled, rammed, and grabbed sediments and rocks from the world's oceans. To complement actual such data, scientists use seismic profiling and sidescan sonar to obtain a larger-scale picture of the seafloor and its underlying structure. In seismic profiling, sound is bounced off the bottom at frequencies that penetrate through soft sediment and reflect or refract off hardened subsurface layers. The returning sound waves are then used to create an image of the subsurface geologic structure. In sidescan sonar, the echoes from a fan of sound pulses are used to image a wide swath of the seafloor's surface. By overlapping the swath images from sidescan sonar data, an aerial photo-like picture of the seafloor is generated. Sidescan sonar also provides information about what the seafloor is composed of. Soft sediments like mud reflect less sound than harder materials such as rock and sand. The very latest technology for sediment and seafloor mapping combines highly accurate depth profiling with the collection of backscatter data (similar to sidescan sonar). In the 1990s, high-resolution multibeam swath-mapping systems were developed to simultaneously collect depth data with unprecedented accuracy, obtain a swath image of the seafloor, and use very precise, differential GPS for accurate locating (Gardner et al., 1999). The U.S. Geological Survey and it collaborators have already produced spectacular new images of the seafloor in Massachusetts Bay, Monterey Bay, Stellwagen Bank, and San Francisco Bay (Figure 39). By combining actual sample data with information obtained indirectly through seismic profiling, sidescan sonar, and

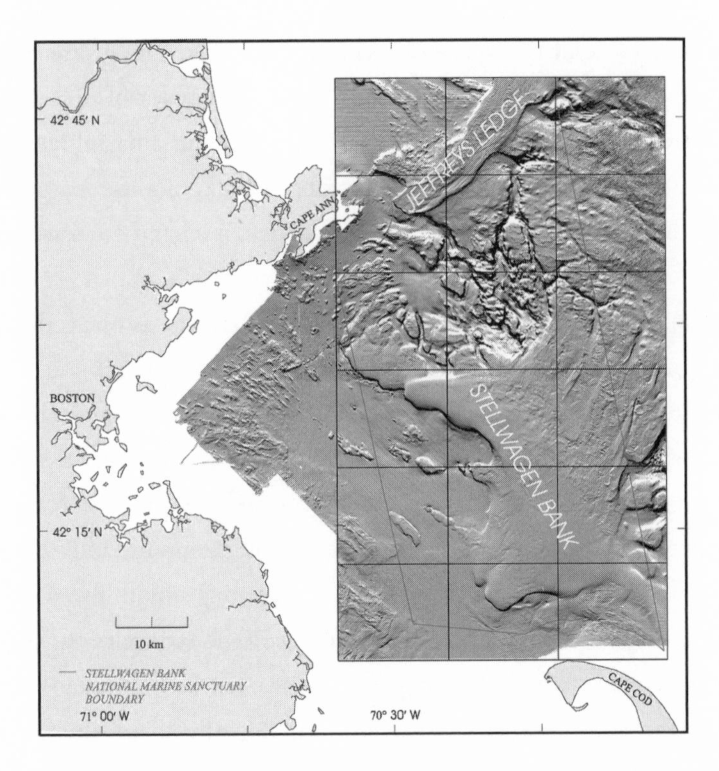

Figure 39 Composite map of Stellwagen Bank made from backscatter and high precision depth data.
(Courtesy of the U.S. Geological Survey.)

backscatter data, maps of marine sediment distribution and the subsurface structure are rapidly improving. But because we have only sampled or observed first-hand a miniscule portion of the seafloor, much of our understanding of the global distribution of the ocean's sediment is an extrapolation of the data or, basically, an educated guess.

The distribution of sediments in the ocean is related to the processes that produce and bring sediments to the sea and seafloor, and those that alter the sediments once they get there. For the purposes of mapping, marine sediments are generally divided in four groups: glacial, terrigenous, siliceous, and calcareous. Glacial sediments, those associated with the frigid grip of ice, tend to accumulate mainly in a broad band of gravel encircling the shores of Antarctica. Other much smaller regions of glacial debris are found in the far north, for instance, just east of Greenland. Terrigenous (land-derived) sediments, as one would expect, rim the continents and are of particular abundance where rivers enter the sea. Siliceous sediments, primarily diatom and radiolarian oozes, occur in three distinct stripes, along the equator and at high latitudes, both north and south. The distribution of silica-rich sediment in the sea reflects mainly the depth and fertility of the overlying water. Radiolarians and diatoms grow in abundance in nutrient-rich waters. Hence, in zones of upwelling such as along the equator, at the coast, and in the southern ocean, great quantities of siliceous shells rain down from above. As the shells sink, some of the silica is dissolved in the surface waters, but as temperature decreases and pressures increase with depth, dissolution wanes. Shells that make it through the surface waters settle on the seabed and become part of the sediment. In deep regions, where productivity is low, red-brown clay coats the seafloor. The distribution of calcium carbonate in deep marine sediments differs from either silica or clay and coincides with the location of the midocean ridges. Calcareous sediments are found principally in two regions of the ocean: in shallow-water regions where modern or ancient reefs and other carbonate buildups occur, and in the deep sea, along the flanks of the midocean ridges. It is the whitish, calcareous oozes that produce the "snow-toppped" peaks of the underwater realm.

Whereas silica tends to dissolve near the surface, calcium carbonate dissolves in the deeper sea. The increase in pressure and decrease in temperature with depth causes calcium carbonate to dissolve. On average, below about 4 to 5 kilometers, almost all calcium carbonate is dissolved. Consequently, only those areas of the seabed that rise above a depth of 4 to 5 kilometers, such as the peaks of undersea mountains, are blanketed by the white of millions of tiny calcium carbonate shells. The level at which complete dissolution of calcium carbonate occurs is shallower in the Pacific than in the Atlantic and has varied over time depending on the ocean's chemistry, circulation, and the amount of carbonate material entering or being produced in the sea.

Unlocking the Secrets of a Deep-Sea Core

When a core of sediment is collected from the deep sea, a rich history of the ocean's biology, geology, chemistry and Earth's climate and plate motion is unlocked. However, deciphering the information laid bare in a sediment core is a tricky business. Not only do conditions within the water and air above the seafloor vary over time, but also because of plate tectonics, the actual position of the seafloor changes. Those studying core samples become geologic detectives and rely on knowledge, experience, imagination, and some simple principles to help decipher the secrets of a deep-sea core.

In a core where the sediment layers are intact and undisturbed, younger sediments overlay older sediments. In other words, the bottom of the core was deposited before the top of the core. If shells or a layer of ash within a core can be dated, usually using radiometric techniques, then the sediments above the dated section are younger and those below, older. It is as if a dated newspaper was thrown into a garbage can; all the waste below would be from the days before and anything on top would be from the days after.

The thickness of a sediment layer is a measure of time and the process that produced it. For instance, a thin layer of volcanic ash represents almost an instant in time, as compared to a slowly settled layer of fine foraminifera ooze. Sediment thickness may also be affected by compaction. The sediments near the bottom of a core will have been compressed more than those near the top. Mixing by marine organisms that crawl through the sediments or create burrows can blur layering and thus the time history of a core's geologic record. Even with the difficulties involved, a geologic history can usually be deduced from the sediments of a typical deep-sea core based on our understanding of plate tectonics, seafloor spreading, and the modern distribution of marine sediments. A sequence of events and the core that they would produce are described below in a typical, though purely hypothetical, example.

At the crest of a midocean ridge, molten rock cools to form the dark, ragged blocks or smooth pillow basalts of the ocean's crust. As the seafloor spreads away from the ridge crest, the dark, volcanic crust ages and cools. Cooling causes the crust to contract, increase in density, and sink into the underlying mantle. The depth of the overlying water increases. In fact, in one of the few simplistic relationships that seem to work, one can calculate an estimated age of the seafloor by its depth.

In the waters overlying the new crust, fine flakes of clay and the small shells of foraminifera rain down. At first, a thin skin of ooze covers the crust in a blanket of white. Over time, the sediment blanket thickens. As long as the seafloor remains shallower than 4 to 5 kilometers, calcium carbonate sediments accumulate and overwhelm the bits of clay. A core taken at this location would have two layers—a base of dark volcanic crust and an overlying layer of light-colored calcium carbonate ooze (Figure 40).

Time marches on and the seafloor continues to spread and sink. Now the seafloor lies in depths of some 6 kilometers, and as calcium carbonate shells fall toward the seabed, they dissolve. Soon the only particles that settle onto

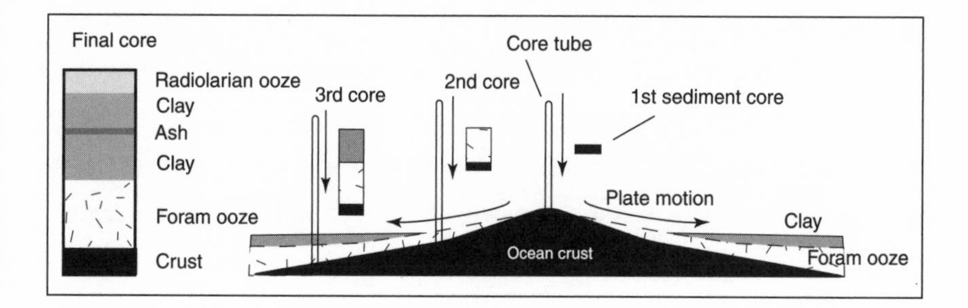

Figure 40 Illustration of hypothetical deep-sea cores.

the seafloor are fine flakes of clay; some encapsulated in an organic sheath from the digestive processes of marine organisms. For some reason seafloor spreading slows and a thick brown layer of clay builds up on the seafloor. A core collected now has three layers: crust, carbonate ooze, and clay.

A colossal volcanic eruption then occurs on an island in the Caribbean and spews an enormous volume of ash skyward. Over the next year a fine blanket of volcanic ash settles into the world's ocean and produces a thin layer on the seafloor. After all the ash settles out, clays again accumulate alone. A core collected at this point would have five layers from bottom up: crust, carbonate ooze, clay, ash, and clay.

Though it has been spreading slowly, the seafloor now nears the coast and enters an area of coastal upwelling. The shells of siliceous organisms rain down in such profuse abundance that they overwhelm the tiny particles of clay and begin to accumulate as radiolarian ooze. One final core is collected and it has six layers, the five previous topped by a grayish deposit of silica-rich sediment (Figure 40). Alas, as time progresses and the seafloor spreads, our piece of the ocean bottom is driven into a trench and subducted into the underlying asthenosphere. Bits of the seafloor and its sediment are scraped off the down-going slab and added to the continental landmass on the other side of the subduction zone. On the subducted plate, increasing heat and pressures within Earth melt the remaining ocean crust and sediment. Some of the melt is recycled into the asthenosphere and some is driven upward through cracks and holes in the overlying rock. And now, beneath a dormant volcano along the ocean's edge, a chamber fills with molten rock. If the volcano were to erupt some years later, who would guess that some of the material in its ash and lava was at one time lying beneath the sea and, at an even earlier moment, was a part of the living, biological ocean.

The Biological Ocean

In the whole ocean, as large as it is, there lives nothing that we do not know.
—Pliny the Elder (23–79 A.D.)

BASED ON OUR knowledge of the seas today, or lack thereof, the above quote should really read: "In the ocean, as large as it is, there lives a whole lot and most of it we know little about." The sea overflows with life, from the tiniest marine microbes to the largest of mammals, from the deep sea to the shallow shore. But our knowledge and understanding of the creatures that live in the sea are poor. For years and with little choice, scientists sampled the oceans by blindly towing a net or dragging a dredge behind a moving ship. Some organisms could evade the towed nets, while others were too large or too small to be captured. Only the blind, stupid, and unlucky of the sea were caught. The unfortunate few ensnared by the sampling gear were dragged along in a swirl of water, organisms, and sediment. Picture a delicate jellyfish, whose body is little more than a thin sheath of tissue over a mass of watery, gelatinous goo, being towed in a net behind a boat. Its delicately designed body is broken, squashed, and jumbled up with the rest of the net's contents, until it is little more than a glob of slime, sometimes fondly referred to as sea snot. Clearly, this method of collection leaves much to be desired, and we learn almost nothing about the behavior of marine creatures when we snatch them from their briney home.

Scientists today use submersibles, scuba gear, and remotely operated vehicles to collect and observe marine creatures, but even these methods have their limitations. Imagine being 30 meters below the surface, hanging off a thin line from a boat at night, and breathing air from a scuba tank strapped to your back. The sea around you is amazingly quiet, mesmerizingly clear, and eerily dark. You see tiny creatures flitting about in the water and transparent plum-shaped organisms swim by in pulsating wisps of color. Every once in a while, though not all that often, a bigger creature swims or floats past. It may be a fish a few centimeters long with glowing blue spots along its body, or it may be a transparent gelatinous creature with tentacles several meters in length. Now you can see how the organism moves in the sea, what it looks like suspended in the ocean's watery flow. But unless you can follow that creature around, you will learn little about its behavior. Does it spend all of its life at one depth or migrate during different times of the day or season? How does it find a suitable mate for reproduction in the vast reaches of the sea? And, quite simply, how does it find food and eat? These are very basic questions, but in the ocean they are extremely difficult to answer.

For the larger creatures of the sea, such as whales, fish, turtles, and sharks, we have acquired a new view into their lives and behavior thanks to the development of sophisticated tagging and satellite tracking techniques, as well as newly designed video cameras called *animal cams.* Tagging and satellite tracking techniques enable researchers and even the Web-savvy public to follow larger creatures, such as whales and turtles, for days, weeks, and even months as they swim through the ocean realm. Animal cams can be attached noninvasively to an animal and allow us a first-person view of life in the sea. Submersibles and remotely operated vehicles enable us to enter and observe the very deepest parts of the ocean, but our stay there is still very much restricted in time and space, and when we try to bring specimens from their home in the deep sea up into our own shallow-water world, the changes in pressure, temperature, and light are often fatal. Today, powerful microscopes provide a view of the ocean's smallest creatures, the bacteria or microbes, and genetic research is beginning to reveal information about the heritage, ancestry, and links within populations of marine organisms. It is an exciting yet still frustrating time to study the biology of the sea.

Life's Diversity in the Sea

We have progressed far in our ability to sample and view the creatures of the sea, but remain hampered by the ocean's vast and inhospitable nature. We still have much to learn about—not only *what* lives in the sea, but maybe more importantly—*how* it lives and what factors control or limit life in the oceans.

The oceans are vast in both space and time, representing the largest living space on the planet today and throughout Earth's history. The creatures that inhabit the wide sea are diverse in both shape and form. Some have the ability to swim against the ocean's flow, while others are relegated to a life of floating amidst the waves. Many of the sea's creatures live within the mud, sand, and rocky bottoms that floor or rim the ocean basins, while others prefer a strictly watery domain. Though most of the ocean's living space is contained within its open waters, only a relatively small percentage of marine life lives there. In the open sea, creatures must find a means to stay afloat, find food, attract mates, and avoid predators in a vast medium with few, if any, hiding places. The following sections provide a brief but intriguing look at the ocean's biological diversity and some of the fascinating survival strategies used by organisms in this watery realm. Unquestionably, the text here cannot do justice to the myriad of evolutionary wonders that nature has constructed in the sea. For greater detail, see the readings listed at the end of the book.

The Drifters

The plankton have evolved some fascinating and often surprising strategies to make the most of an existence floating among the waves. The term *plankton* comes from the Greek word *planktos* meaning "drifting" and is generally used to describe the free-floating or weakly swimming organisms of the sea. Though we often distinguish between phytoplankton and zooplankton as the ocean's floating flora and fauna, the division is somewhat murky. Some of the plankton, like the dinoflagellates, have both plant and animal traits; they have special chlorophyll-containing cells for photosynthesis, but they are also capable of capturing and ingesting prey. This gives them the extraordinary ability to sustain themselves in varying circumstances: when light levels are sufficient they can use photosynthesis to derive energy from the sun, but when light levels are too low, they can hunt for their food.

Plankton are single-celled or multicelled organisms and live as individuals, as colonies, or in some cases, as part of a multiorganism partnership. Some plankton are simple rounded spheres, while others have a more elaborate and ornate form. Many are truly passive floaters, while others, though not strong enough to swim against currents, are able to move ever so slightly within the ocean's unceasing flow. Some marine organisms, such as the lobster, eel, coral, and starfish, are plankton for only part of their life, spending their youth floating with the currents and their adulthood on the bottom or moving about as true swimmers. The planktonic lobster larva is a strange-looking thing, more like a

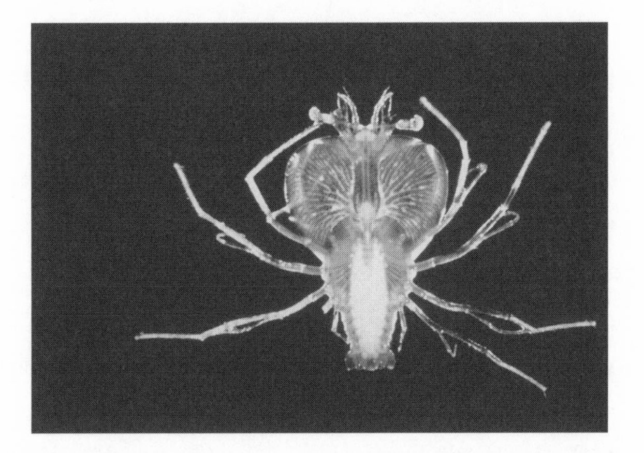

Figure 41 Odd-looking skeleton-like lobster larva, a temporary member of the plankton.
(Courtesy of Florida Fish and Wildlife Conservation Commission, Florida Marine Research Institute.)

wafer-thin skeleton than a many-legged, thickly shelled crustacean (Figure 41). It may float among the plankton for days, weeks, or even months before settling to the seafloor and growing into the more familiar lobster form. Because many coral, fish, and crustaceans have planktonic larvae, ocean currents play a major role in determining where adult populations exist. Some of Florida's lobsters and fish may actually originate upcurrent off the coasts of faraway islands in the Caribbean.

For its inhabitants, the ocean is an eternal hunting ground, and size plays a major role in the life-long game of eat or be eaten. How big you are often determines what you can eat and what can eat you. The smallest, most vulnerable—and yet quite possibly the most important—creatures of the sea are the floating marine plants and algae, the phytoplankton. The phytoplankton are at the base of the ocean's sun-driven food web (to differentiate from the chemical-driven web at deep-sea vents). Through photosynthesis, they produce organic matter from the sea's inorganic materials, this is often called *primary production*.

Phytoplankton

The phytoplankton are generally small (less than 1 millimeter in size), rounded, single-celled creatures that are well suited for life in the open sea. Their small size and spherical shape result in a high surface area to volume ratio and consequently a high floatability. This is crucial, because to photosynthesize and grow, the ocean's plants and algae must stay within the sunlit surface waters. To keep from sinking into the sea's darkness, phytoplankton remain small; they may also have small whiplike tails or increase their surface area with spines or by forming chains with their neighbors. With no mouth, stomach, or circulatory system, these single-celled plants and algae rely on simple diffusion to absorb nutrients and release wastes.

Diatoms and dinoflagellates are the most common phytoplankton in the sea. The diatoms, single-celled algae with glassy silica shells, commonly group together into long chains. Compared to some of the other phytoplankton, diatoms can be rather large—even visible to the naked human eye. The shell of the diatom is shaped like a pillbox, with tightly fitted top and bottom lids made of silicon dioxide, the same material that is the major constituent of glass. They have no known means of locomotion, though some have spines or spikes to help prevent sinking. The diatoms frequently reproduce through simple asexual division; each diatom divides by replicating a smaller cell in the top and bottom half of its original shell. Consequently, over time and with each new generation, individual diatoms get smaller and smaller. Eventually one of the increasingly small diatoms casts off both portions of its shell and creates a special type of cell that can reproduce a larger-sized diatom. When conditions become harsh, some of the diatoms can also form nonreproducing resting spores. When more favorable conditions develop, the resting spore once again becomes a viable and active diatom. Diatoms are abundant in the ocean's nutrient-rich regions, particularly in cold zones of upwelling along the coasts and in the polar regions.

Dinoflagellates are small, free-living marine creatures that overcome sinking with the use of their two small tails called flagella (Figure 42). They are an extremely diverse group of phytoplankton that come in a wide variety of shapes and sizes. Some have shells made of chitin (a key component in the shells of shrimp, crabs, and other crustaceans), while others live perilously naked. Dinoflagellates can be considered both plant and animal; some have chlorophyll and others do not. Those without chlorophyll always behave "animal-like" and engulf food particles, much like an amoeba. If conditions are right, dinoflagellates can reproduce rapidly and in great abundance, creating an ocean bloom. Certain dinoflagellates can produce a chemical form of light called bioluminescence (discussed below) they create magical nighttime displays like sparkling undersea stars. Other dinoflagellates are more menacing, causing red tide, shellfish and ciguatera poisoning. Some of the dinoflagellates can form cysts that lie temporarily dormant in the sediments for long periods of time or live in partnership with other organisms.

The ocean's phytoplankton also include the coccolithophores. They are brown algae, have flagella, and are covered by a spherical sheath of microscopic calcareous disks or shields (see Figure 11). Blooms of the tiny coccolithophores can cause the sea surface to turn milky white, and extremely dense concentrations can be detected on satellite images of ocean color. Coccolithophores are most common in the tropical and warm-temperate seas.

Cyanobacteria, formerly called blue-green algae, are an important, though poorly understood, constituent of the plankton. Some cyanobacteria photosynthesize, while others perform a feat rare in the ocean realm: they use nitrogen in its free molecular form. Most plants cannot

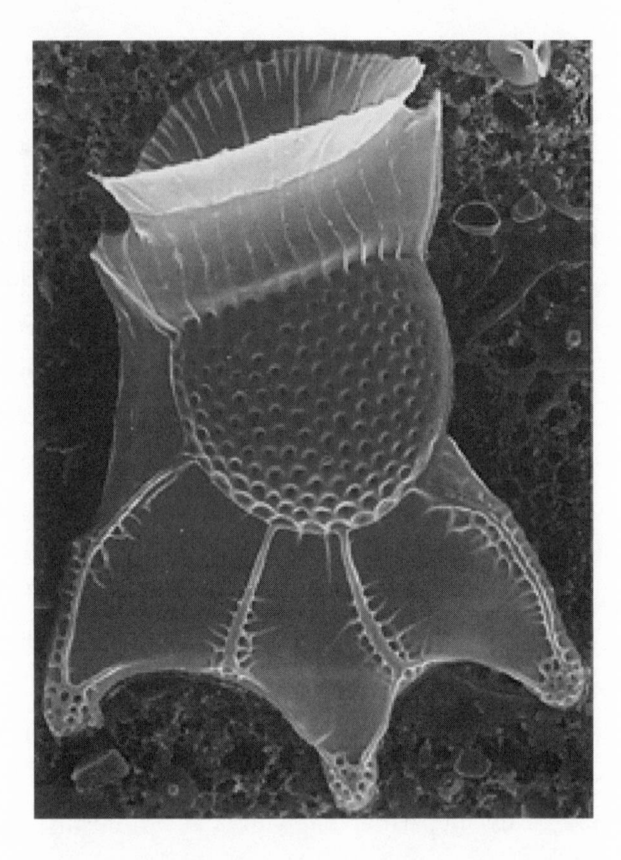

Figure 42 Photograph of a dinoflagellate under a scanning electron microscope.
(Courtesy of Vita Pariente, Texas A&M University.)

use free molecular nitrogen, but require an altered form, either nitrate or nitrite. By converting free nitrogen in the water into more usable, plant-friendly forms, cyanobacteria play an important role in the sea, and are often called "nitrogen fixers." They are particularly abundant in nutrient-poor regions like the tropics. One type of cyanobacteria, found in warm, tropical waters, is the harmless *trichodesmium,* which forms small, floating brown tufts and puffs. Sometimes the water is visibly full of these spindly or rounded bacterial bundles. Many other types of bacteria live within the ocean either as free-floaters or as bottom dwellers. They are very important in both the production and decomposition of the sea's organic matter and in the recycling of nutrients. Because bacteria are so small, they were typically overlooked by traditional sampling techniques; more modern methods allow for a better view of the bacterial world. Most scientists believe that bacteria and other microbes in the sea are even more plentiful and significant than has already been found.

The distribution and abundance of phytoplankton is strongly controlled by the sea's temperature and the availability of light and nutrients. In areas of the ocean where nutrients are particularly lacking, the smaller phytoplankton dominate. In nutrient-rich areas, like upwelling zones, the larger species proliferate. And unquestionably, where the phytoplankton prosper, so do the zooplankton.

Zooplankton

Zooplankton are the drifting animals of the sea whose watery activities include the hunting and eating of others. Collectively, zooplankton are considered the ocean's most important grazers of phytoplankton; thus, they are the link between primary production and the growth of larger life-forms. The zooplankton are more diverse than the ocean's floating plants and grow much larger. On land, plants can grow larger than animals—compare the great sequoia with the comparatively dainty African elephant—but in the sea, animals usually grow larger than plants. This is partly because water is more dense than air and thus allows larger growth with less of a support structure. The smallest of the zooplankton are the single-celled creatures, such as the foraminifera, radiolarians, zooflagellates, and ciliates.

Foraminifera are small amoebae encased in tiny shells of calcium carbonate (see Figure 14). They tend to be less than 1 millimeter in diameter, but can reach a centimeter or so in size. As foraminifera grow, they add chambers to their shells. Most foraminifera feed on bacteria and phytoplankton by extending gooey, sticky arms of protoplasm, called pseudopods, out through pores in their shells and engulfing prey. Some foraminifera also have algae living within their tissues that aid growth by removing wastes and producing oxygen. Foraminifera are common throughout the world's oceans both as planktonic organisms and as benthic creatures, living on or within the seafloor sediments.

The radiolarians are a group of relatively small spherical zooplankton with elaborately decorated shells of silica. Single organisms are about 1 millimeter in size, but colonies can reach several centimeters in length. Like the foraminifera, the radiolarians typically eat bacteria and phytoplankton; however, they capture their prey with gooey tissue that is external to their shell. They are common in the warm, nutrient-rich waters of the equatorial upwelling zone. Some of the smallest, least known zooplankton have flagella (like the dinoflagellates) and are called zooflagellates, and some have cilia and are called ciliates. Cilia are short, hairlike extensions that can beat in concert and produce a vibrating motion that is used for feeding or movement.

Many of the small animals and plants of the sea are eaten by the larger, multicellular zooplankton, such as the jellyfish, copepods, krill, shrimp, some shellfish, small squid, and the voracious but diminutive arrowworms. Jellyfish fall into a group of organisms commonly referred to as the gelatinous zooplankton, those with bodies of watery, viscous goo. Within the ocean, having a soft, transparent body made of 95 to 98 percent water can be advantageous. It is difficult to see, has low nutrient requirements, stays suspended in the water relatively easily (having a density similar to that of the surrounding water), and can easily engulf food particles.

Essentially, a jellyfish and its close relatives are simply a big stomach surrounded by a ring of tentacles—and they all feed on meat. The tentacles contain specialized stinging cells—a sort of miniaturized spring-and-harpoon system—which are used to capture, paralyze, and move prey into the gut. In some species, the "harpoon," or barbed thread, delivers a paralyzing, or even lethal, dose of toxin. As some people know all too well, the potency of the stinging cells varies from species to species. The common round-moon jellyfish with its umbrella-like swimming bell and relatively short tentacles has weak stinging cells. However, its cousin, the box jellyfish, can pack a powerful punch and inflict a painful and dangerous sting. The box jellyfish has a small, cube-shaped swimming bell, less than 15 centimeters in length, with clusters of long, thin tentacles extending from each of the lower corners. Probably the most dangerous of the jellyfish are the Pacific sea wasps, found off the coasts of Australia and the Philippines. An encounter with a sea wasp can leave a person permanently scarred, seriously ill, or in some cases, mortally wounded. Corals are related to the jellyfish and contain similar stinging cells, though only the fire coral is truly potent.

Jellyfish move through the water by means of a rhythmic pumping of their gut. Some pump their way to the surface and then slowly sink, capturing the unwary and unfortunate on the way down. They then pump their way back up and repeat the process. One type of relatively harmless yellow-brown jellyfish, *Cassiopeia,* spends most of its time upside-down on the bottom. It is very common throughout the tropics and can often be seen in shallow water around docks. Sometimes a lobster larva hitches a ride on top of a jellyfish's swimming bell. It is thought that the young lobster relies on the tentacled defense of its pulsating carriage to ward off predators.

A gelatinous cousin of the jellyfish is the siphonophore. One of the most common siphonophores is the dangerous Portuguese man-of-war. It has a purplish-blue tinted balloon at the surface and long, skinny, and painfully stinging, blue tentacles below. Although siphonophores look very much like an individual jellyfish, they are actually floating colonies. Individual organisms form units within the colony, each having different responsibilities. Some creatures make up the tentacles hanging below the gas-filled float; others, the pumpers, are responsible for movement; and still another group of individuals forms the mouth and stomach. Once digested, nutritive materials are diffused through the communal body cavity to all the individuals in the colony. Scientists studying siphonophores in the deep sea have encountered colonies reaching up to 20 meters in length (Harbison, 1992), though most are smaller. The siphonophore aptly nicknamed by-the-wind sailor is a small, harmless, bluish creature with a flat, hornlike float, or "sail," that projects up into the wind at the sea surface. The floats of both the Portuguese man-of-war and by-the-wind sailor allow them to stay near the food-rich surface, but also put these creatures at the mercy of the wind and currents.

Comb jellies, also called ctenophores, look similar to jellyfish; they have a transparent plum- or walnut-shaped gelatinous body with short tentacles or none at all. They are smooth, transparent organisms with eight rows of tiny hairs (cilia) along their sides. The hairs beat in unison and create a minor form of propulsion. Comb jellies with tentacles typically do not have stinging cells. All of these graceful creatures are bioluminescent and can create colorful displays of light (Color Plate 12). They are also voracious carnivores, capturing food particles with sticky tentacles or engulfing them with their oversized mouths. Large aggregations of comb jellies can turn the water into a mass with the consistency of Jell-O. Swimming through this thick soup is not dangerous, but rather unpleasant.

The presence of gelatinous animals in the sea has been known since the first mass of jelly washed ashore or slimed a fisherman's net. But until recently we knew little about their appearance in their natural environment, and almost nothing about their behavior. Our understanding of gelatinous creatures came from chance sightings and from specimens collected using the towed-net strategy, which churned the jellies like delicate lingerie in a violent washing machine. Undoubtedly, gelatinous zooplankton have been some of the most misrepresented creatures thanks in large part to destructive net collections. For example, the communal siphonophores were once believed to be small, individual creatures because in so many instances their large, delicate colonies were broken up into small, transparent pieces during collection. The old way of studying these creatures was akin to examining butterflies by throwing a handful of the winged creatures into a blender, tossing in a few beetles, and hitting "puree"; after which you would separate out the mashed and mangled bodies and then painstakingly try to piece them back together again in the hopes of learning something about the original organisms. Marine scientists are now using scuba equipment, submersibles, and remotely operated vehicles to study the gelatinous zooplankton. It is a slow, tedious, and often frustrating task to hang out and wait for a jellyfish or siphonophore to pulsate by, but it is undoubtedly worth it. These methods are revealing some of the sea's most delicate and beautiful creatures (Color Plate 12).

Although we think of most shellfish as bottom dwellers, there are a few varieties that live in the open water and are considered zooplankton. The pteropods, or sea butterflies, are weakly swimming, upside-down snails. Most have small, fragile, tubular shells of calcium carbonate and a winged foot for paddling. Relatively recent research has revealed that the pteropods use a sticky net of mucus to collect and feed on particulate matter in the water, a process likened to blowing a huge bubble of chewing gum in a room full of insects and then slurping up the gum and all that it captures. Naked pteropods tend to be swifter swimmers than the shelled variety, and they feed upon their slower relatives. A thick shell can be beneficial for protection, but in the water column, it can hinder swimming and promote sinking. One tiny snail with a small-coiled shell, called *Janthina,* has found an ingenious solution to the sinking problem: it creates and attaches itself to a raft of bubbles that floats at the surface.

Unquestionably, the most prolific zooplankton in the world's oceans are the copepods. Their great abundance and diversity make them the "insects of the sea." Because copepods graze on the sea's pastures of floating plants, it may however, be more appropriate to call them the cows of the sea or the ocean's livestock. They play a vital role in the marine ecosystem as a link between plant production and upper-level carnivores. Though great in significance, the copepods are small in size, generally only millimeters in length. Most live in the water column, though some exist on the ocean's floor and others are parasitic, living inside the bodies of other marine organisms. Like other crustaceans, the copepods have a rigid exterior carapace of chitin, a segmented body, and numerous jointed limbs. They use their limbs and two long antennae to move through the water in a kind of jerky sink-swim motion. The copepods are extremely efficient and selective grazers, either filtering or capturing their prey. A modified set of legs can beat up to 2000 times a minute to circulate water through their appendages and create a highly effective food-filtering system. Copepods feed mainly on phytoplankton and other bits of floating particulate matter. They excrete their wastes as consolidated pellets, which play an important role in nutrient recycling and particle sinking in the sea. Many creatures of the sea consider copepod fecal material fine dining; some suggest that the lure dangling from the head of the deep-sea anglerfish is meant to resemble a tempting copepod fecal pellet. Brings a whole new meaning to the comment, you look like crap today.

Another of the zooplankton crustaceans is the krill. They look similar to copepods, but are compressed lengthwise, have a loose-fitting carapace, and tend to be larger, reaching lengths of some 1 to 5 centimeters. Krill are particularly abundant in cold water at high latitudes and are the favorite food of some of the ocean's most charismatic megafauna, the whales.

It has been suggested that another of the crustacean zooplankton, the amphipods, were the original inspiration for the hideous creatures in the sci-fi movie *Alien*. Amphipods can look shrimplike, or they can be rather unique in form, in a grotesque sort of way. A mass of thin, long, jointed legs, claws, an overly large head, and bulging, beady compound eyes makes this undersea crustacean a scary but, luckily, small threat. Amphipods scavenge and feed on particulate matter or smaller zooplankton. Some rather nasty forms are parasitic, preying on others such as the salps, which are simple, transparent, barrel-shaped gelatinous creatures. A female amphipod will kill a salp, devour the internal organs, and save the outer casing as an incubator for her eggs—a devious plan indeed.

Chaetognaths, commonly called arrowworms, are also a frequent constituent of the sea's floating assemblage of animals. They are transparent, ranging from a few millimeters to several centimeters in length, and have fins for swimming. Although they have no eyes, arrowworms are aggressive predators, using cilia to sense water movement and movable bristles about their mouths to seize their prey. If they were larger, the arrowworms would undoubtedly pose a serious risk to all that live in or visit the ocean. Luckily, many of the sea's most voracious predators are very small, invisible to human eyes. For us, the ocean is a relatively benign and often serene place, but for most of the sea's creatures it is a perilous world.

The Daily Migration

Each day at dusk, a great vertical migration occurs in the sea. As the light begins to fade, zooplankton and small fish that live relatively deep migrate upward into shallower waters. As they rise toward the surface, these creatures cover a distance equivalent to some 50,000 times their body length; in human terms that equals a daily walking commute of some 80 kilometers!

The first clues to the presence of the sea's vertically migrating population arose when people began probing the ocean with sound to measure depth, hunt for submarines, and search for schools of fish. The acoustic waves from an echo-sounder or Fathometer would reflect off a mysteriously moving midwater layer, a sort of false bottom suspended between the surface and seafloor. This strange "phantom bottom" was detected at some 300 to 500 meters during the day and then at shallower depths, less than 100 meters, at night. It was found throughout the world's oceans and became known as the deep scattering layer (DSL).

At first, the cause of the DSL was thought to be hordes of plankton or a great school of squid and fish. Then people began to suspect that organisms with gas-filled floats, such as the siphonophores and fish with swim bladders, were the cause. Both the swim bladder and a gas-filled float are believed to effectively scatter the sound waves produced by an echo-sounder. The swim bladder is an organ in many of the bony fishes used to control buoyancy. It is located in a fish's top half to prevent it from accidentally flipping upside down and is controlled by a special gland that can either secrete or absorb gas into the bladder. If a deep-water fish is brought to the surface too quickly, often its gas gland cannot adjust fast enough; and it overinflates. Submersible observations and net collections have now revealed that a variety of creatures make up the DSL. So far, few squid have been found, but notable aggregations of krill and shrimp have been documented. It appears that organisms within the DSL may form layers, segregated by species (Ellis, 1996). Probably the most abundant organism in the sound-scattering thicket—and in the sea—are the tiny lantern fish, or myctophids.

Lantern fish range in size from about 5 to 15 centimeters; they are dark or silvery in color and have relatively large heads and eyes. Their name stems from the dotted rows and patches of photophores that decorate their sides. *Photophores* are small, modified mucus glands used to produce a glowing blue light. The arrangement of photophores, sometimes numbering 50 to 80 along the fish's head, belly, and sides, is used to distinguish different species. Some species of lantern fish roam across the entire width of the ocean, while others stay within a restricted range of water temperatures or within a particular mass. Observers have noted thick clouds of lantern fish at depth during daytime submersible dives, their bioluminescence creating a remarkable display of flashing blue light. As sunset approaches, the lantern fish migrate upward toward the sea surface; when the sun rises, they and the other members of the DSL begin to descend once again into the sea's deeper, darker realm.

Why this mass vertical migration occurs on a daily basis is poorly understood, but several explanations have been offered. Throughout the oceans, most of the food is at or near the surface, particularly the tasty organic-rich phytoplankton. But during the day, when the sun is high and temperatures are warm, the surface is a dangerous place for creatures to linger. It is easy to be seen by predators in the light of day. To take advantage of the wealth of food at the surface, but to avoid being eaten, the creatures of the DSL stay in the dark, deep waters of the sea during the daytime and then, under the cover of night, stealthily migrate to the surface to feed. Most fish and zooplankton are cold-blooded, so the surrounding water temperature regulates their own body temperature and metabolism. By staying in deeper, cooler water by day, they conserve their energy by slowing their metabolism. At nighttime, the surface waters are relatively cooler, and organisms may migrate upward to refuel efficiently; in other words, they expend less energy while feeding.

The vertical migration of the deep scattering layer is cued by changing light levels. On dark nights, when the moon is new or the sky cloudy, vertical migration is quite pronounced. In contrast, on bright nights when the moon is full and the sky is clear, vertical migration may be weak or absent. The sea's great vertical commute is most conspicuous in the tropics and essentially nonexistent at the poles. This is because the sharpest temperature and daily light gradients occur in the tropics, while at the poles, there is little temperature variation with depth and less of a daily change in light levels.

Bioluminescence

At night along the sea's edge, the ocean sometimes seems to glow, as if lit from within. The foamy wake behind a boat can shine brightly in the dark of night, and a sparkling twisty trail of light sometimes announces the nighttime antics of a dolphin or sea lion. These observations are not the fodder of myth or legend, but the result of bioluminescence; a phenomenon exhibited by many of the sea's zooplankton. Bioluminescence is the production of cold light through biological processes, as opposed to phosphorescence or fluorescence, both of which are re-emitted light that was initially absorbed from an external source.

Many of the sea's creatures, including squid, dinoflagellates, bacteria, worms, crustaceans, and fish, are known to produce light. The process that marine creatures use to create light is like that of the common firefly, and similar to that which creates a luminous green color in the plastic glow-sticks that are so popular during nighttime events. When a glow-stick is bent, two chemicals mix, react, and create a third substance that gives off light. Bioluminescent organisms do essentially the same thing; they have a substance, called luciferin, that reacts with oxygen in the presence of an enzyme, luciferase. When the reaction is complete a new molecule is formed that gives off light—glowing blue-green in the underwater world. This biologically driven chemical reaction occurs within the organism's special light-producing cells, photocytes, or organs, photophores. In some organisms the photophores are simple glandular cups; in others they are more elaborate devices with lenses for focusing, a color filter, or an adjustable flap—sort of an on-off switch. Probably one of the most complex light-producing systems is that of the squid. Some squid have both photophores and chromatophores (organs for changing color) within their skin, thus enabling them to control both the color and intensity of the light produced. Recent research has also revealed that in some squid and fish, bioluminescent light may be produced by bacteria that live in a mutually beneficial partnership inside the animal's light organs.

How and why bioluminescence occurs is not fully understood; however, in the undersea realm, it appears to be used in a variety of interesting and ingenious ways. The most commonly observed form of bioluminescence in the sea is the pinpoint sparkling of light at night that can create comet-like trails behind moving objects. Almost always, this is the result of dinoflagellates reacting to water motion. The relatively short, momentary displays of light may have evolved to startle, distract, or frighten would-be predators. Collection nets brought up from the sea's depths at night frequently glow green at great distances. Slowly fading green blobs or pulses of light can be seen coming from the organisms ensnared within, often from gelatinous creatures. This type of light display may be used to stun, disorient, or lure prey. Like the wide-eyed deer on a road dazed by headlights, undersea creatures living within the ocean's darkness may be momentarily disoriented by short flashes of bioluminescent light. Another of the sea's light-producing organisms is a small copepod named *Sapphrina iris.* In the water, *Sapphrina* creates short flashes of a remarkably rich, azure blue light. But its appearance under a microscope is even more spectacular: the living copepod appears as if constructed of delicately handcrafted, multicolored pieces of stained glass. Within the deep sea, some fish also have a dangling bioluminescent lure or a patch of luminescent skin near the mouth, which may be used to tantalize unsuspecting prey.

Other sea creatures have both light-sensing and light-producing organs. These creatures are thought to use bioluminescence as a form of communication or as a means of identifying an appropriate mate. In the lantern fish, the pattern of photophores distinguishes one species from another. In other fish, bioluminescence may help to differentiate males from females. The squid uses light as a means of camouflage. By producing light from the photophores on its underside, the squid can match light from above and become nearly invisible to predators looking up from below. Squid, as well as some of the gelatinous zooplankton, have also been known to release luminescent clouds or strands of organic material, possibly as a decoy to facilitate escape. And finally, because what they eat is often bioluminescent, many of the transparent deep-sea creatures have red or black stomachs to hide the potentially flashing contents of ingested bioluminescent creatures. Without such a blacked-out stomach, their digestive organs would flash like a neon sign that says "Eat me, eat me."

The Swimmers

Organisms that can actively swim against ocean currents are called *nekton,* the true swimmers of the sea. They can effectively battle against currents, control their vertical position, and in many instances, undertake long migrations. The most notable of the nekton are the fish, squid, marine mammals, and sea turtles. Some shrimp and diving seabirds are also considered among the ocean's swimming elite.

The swimmers of the sea occupy three very different environments in the ocean: the shallow, midwater, and deep zones. Within each of these regions, conditions often control how the respective inhabitants behave and look. The shallow zone extends from the surface to about 200 meters in depth. Here, there is ample light, warm water, an abundance of food, and potentially varying conditions. Deeper in the ocean, in the midwater or twilight zone, little light from the surface is available, food becomes more limited, it is colder, environmental conditions are more stable, and many of the inhabitants migrate upward to feed. In the deepest region of the sea, below about 1000 meters, there is perpetual blackness, conditions are constant for the most part, and food is usually scarce, except on rare occasions and at deep-sea vents.

The Shallow Sea

The shallow sea is heavily populated by the plankton, but unquestionably ruled by the nekton. Plenty of food is available to fuel muscles for swimming, and warm temperatures allow for a fast metabolism. With plenty of light available, predators with good eyesight can hunt visually. However, with this wealth of sunlight, no place to hide, and an abundance of visual predators, the shallow sea is a particularly perilous place, so many of the fish that live or feed in this zone have evolved their own specially designed camouflage clothing—a coloration pattern known as countershading. Tuna, wahoo, dolphinfish, and many others of this realm are dark on top, light on the bottom, and sometimes silvery on the sides. Predators looking down may not see the top of a darkly colored fish that blends in with the shadows below; predators looking up into the sunlight may also miss a fish's light-colored underbelly; and when viewed from the side, a fish's silvery scales may reflect the sunlight and present a confusing target.

> *If we were to design the optimum form of life on this ocean planet, it might look remarkably like the majestic bluefin tuna.*
>
> —*Sylvia Earle*

Tuna and other free-swimming fish are also known to aggregate around floating objects in the sea. The explanation for this behavior is unclear; however, given the lack of hiding places in the open ocean, fish may use the flotsam for a temporary reprieve from the ever-watchful eyes of predators. Some fish like the sargassum fish described earlier, have growths and coloration that render them almost indistinguishable from their surroundings. In murky coastal waters, fish tend to be relatively drab in color to meld into the background. Fish that live in the clear waters of coral reefs tend to be very colorful to either blend in with the environment (Color Plate 15), send a warning signal, or trick potential predators. But do fish perceive color the same as we do? Rather than actual colors, most fish may use their sense of sight to detect changes in light intensity and contrast.

The shark is often called the ultimate eating machine, but the true master of the shallow open-water realm is the tuna. While the tuna illustrates many of the characteristics of the other fish within the shallow sea, nature has perfected its swimming system in a streamlined mass of muscle, fin, and scale (Figure 43). The tuna is a member of the incredibly diverse group known as the bony fishes. In contrast to the cartilaginous sharks and rays, the bony fishes have scales, a flap that covers the gills, and movable rays in the fin and tail. As their name suggests, they also have bony skeletons for support. Most have teeth and all have paired fins that are strengthened by rays. The fin rays can be either soft, jointed, or stiff and can be folded flat against the body for better streamlining. In the tuna and other fast-swimming fish, the design is even better than that of the average bony fish; their fins can be folded into specialized grooves in the body to further improve hydrodynamic efficiency. With this and its other

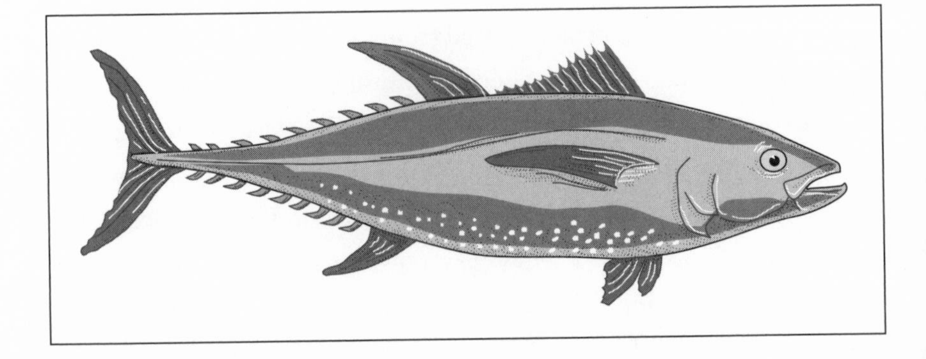

Figure 43 Illustration of the awesome tuna.

speed-enhancing traits, the tuna can swim amazingly fast, at speeds reaching up to 45 kilometers per hour.

The shape of the tuna is the most hydrodynamic shape possible, and one that humans have sought to emulate in submarines and torpedoes. Some call the tuna torpedo-shaped, but really the torpedo is tuna-shaped. The tuna's body is sleek, stiff, and streamlined; it is widest about one-third of the way behind the snout and tapers gently back to the tail. Some of the other billfish and fast-swimming species have a similar form, and all of the bony fishes are bilaterally symmetrical (the left and right sides look the same). The slower-swimming fish have evolved a wide variety of other shapes: long and skinny (needlefish), snakelike (eels), or globe-shaped (puffer fish), flattened (butterfly fish, flounder), or boxy (cowfish). (See Color Plate 15.) Some fish start out tuna-shaped and then change form as they grow. One of the more interesting examples of this metamorphosis occurs in the flatfish, such as the flounder or sole (Figure 44). After hatching from an egg near the surface, the flatfish begins its life tuna-shaped and bilaterally symmetrical. As the young fish grows, its body thins and becomes laterally compressed; one of its eyes moves across its body, over the top of its head and settles close to the eye on the opposite side. At one stage of the process, the fish is completely blind on

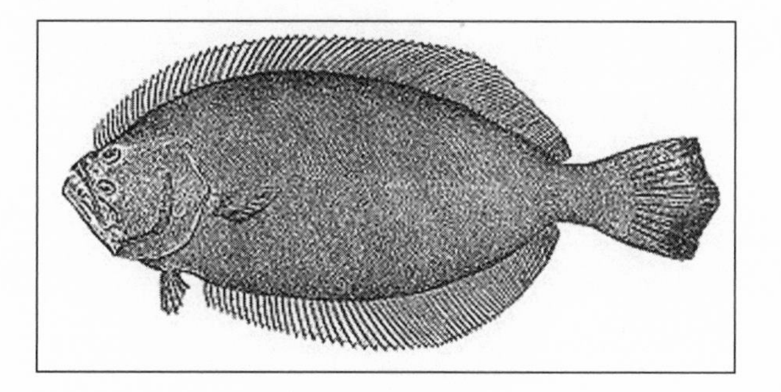

Figure 44 *An adult flounder.*
(Courtesy of National Marine Fisheries Service, NOAA.)

one side. However, once the fish settles to a life on the bottom, the blind side becomes the underside of its body (Waller, 1996).

The tuna's retractable fin gear on its dorsal (top) surface not only enhances its swimming speed but can be raised for better maneuvering in close quarters and to reduce rolling at slow speeds. Behind the dorsal and anal fins, along its top and bottom rear end, the tuna also has a row of small sail-like fins, known as finlets. Researchers suspect that by creating slight turbulence, the finlets reduce drag, much like the drag-reducing dimples on a golf ball (Whynott, 1999). In the slower-swimming fishes, broad pectoral fins are used for braking, and pelvic fins can be positioned to adjust for lift. Departure from the tuna shape and modified fins allow some fish greater maneuverability. In others, highly modified dorsal, pectoral, and pelvic fins are used in a sculling motion; this enables the fish to swim in any direction, including backward (trunkfish, porcupine fish, puffer fish, seahorses, and triggerfish) (Figure 45).

The tuna's tail is another model of exquisite efficiency and hydrodynamic force. Powerful thrust is provided as the tuna moves its stiff, forked, crescent-shaped tail from side to side. At the base of the tuna tail,

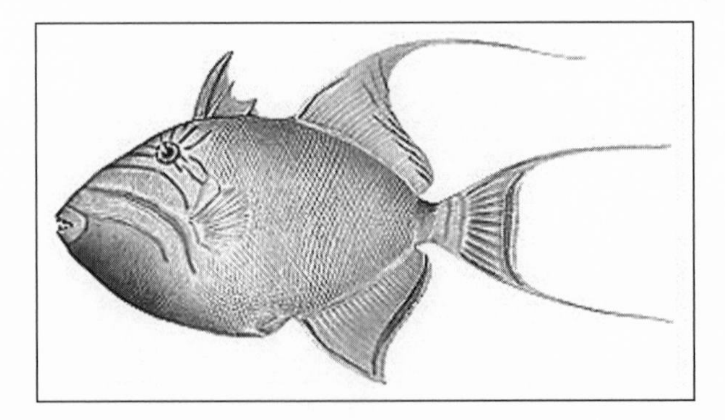

Figure 45 The triggerfish.
(Courtesy of National Marine Fisheries Service, NOAA.)

are two peduncle keels, also thought to reduce turbulence at the tips of the tail fin and lower the drag created by its movement. In slower-swimming and bottom-dwelling fish, tails tend to be broader, flattened, more flexible, and rounder for better maneuverability and less power. The tail and fins on an eel have evolved to produce a highly efficient serpentine motion. In seahorses, the tail has lost all role in swimming; it is more like a monkey's prehensile tail, used to coil around and grip objects such as seagrass, algae, or corals.

Tuna not only swim fast, but they can also cover great distances over short periods of time or during long migrations. At night, tuna are known to cover some 15 kilometers during long offshore excursions (possibly for feeding). In one study of tuna migrations, scientists found that a well-traveled fish had covered nearly 10,000 kilometers from the Bahamas to Norway in a period of just 50 days (Whynott, 1999). For swimming at such great speeds and over long distances, the tuna has also evolved a circulatory system and muscle structure that most athletes can only dream of.

In general, fish move by a series of muscle contractions that begin at the head and progress through the body toward the thruster, the tail. Two types of muscle, white and red, are used in swimming and are arranged along the flanks of the fish. White muscle contracts quickly and is used for bursts of high-speed swimming. It operates anaerobically, generating lactic acid as a by-product (similar to the lactic acid that can build up in humans after prolonged vigorous exercise and produce fatigue or muscle cramps). Red muscle contracts more slowly, is used for slow, continuous swimming, and operates through an aerobic metabolic pathway. Because red muscle does not produce lactic acid, it can continue to contract almost indefinitely with the necessary fuel (oxygen and glucose). In most fish, the bulk of the muscle is white, with only a small lateral strip of red (obvious in fillets or steaks). The tuna, however, is equipped with a much larger proportion of red muscle than the average fish. This allows the tuna to cruise at high speeds aerobically, without lactic acid buildup, or as an athlete would say, without going anaerobic. White muscle is used to generate additional thrust, to go from cruising to warp speed. To sustain such athletic prowess, the tuna requires an efficient means of supplying blood and oxygen to its hard-working muscles, and a way of releasing the heat generated during muscle contractions.

Fish, like other animals, need oxygen to respire and to fuel their working muscles. Dissolved oxygen is some 30 times as dilute in the sea as in the air. Fish use their gills to extract oxygen from the water by pumping water in through their mouths, over and out through the gills. The pumping is done by repetitively opening and closing the mouth, thus sucking or gulping the water in and pushing it past the gills. The tuna's skull and jaw are rigid to enhance swimming speed, so it cannot physically pump water in. Instead, it breathes by means of *ram ventilation*. As a tuna swims, its mouth stays open and water is literally rammed in and over its extraordinarily large gills. Like some of the other fast-swimming fish, the tuna must constantly move or else it will suffocate.

Tuna and some sharks have evolved a specialized circulatory system that uses excess heat production to stay warm in their often-chilly medium. Fish are typically cold-blooded; their body temperature is regulated by the surrounding water. But the tuna has a special internal heat exchanger made up of intermingling arteries and veins. Cool, oxygen-rich blood flows from the gills to the tissues through arteries and a network of small blood vessels. Running parallel to and intermingling with the arterial network is the venous system. In the venous system, warm, oxygen-depleted blood is carried from the organs and tissues back to the gills. Because the veins and arteries are intermingling and in close contact, heat is readily transferred from the warm blood in the veins to the cold blood in the arteries. With elegant physiological simplicity, this countercurrent heat exchange uses the tuna's own internal heat from muscle contractions to warm incoming blood. In most other fish, heat is lost directly from the gills; but in the tuna, there is little heat loss, and by the time the blood reaches the tuna's interior, it is nearly as warm as the internal body tissues. This exchange is so effective that a tuna's core temperature has been measured at 10 to 20°C warmer than the surrounding water. Scientists suspect that elevated body temperatures help keep the tuna's metabolism high for food processing, allows its red muscle to contract more quickly, and may enhance lactic-acid breakdown. Tuna may also be able to partially shut down the heat exchanger if they become too warm, or they may make excursions into deeper, colder water to avoid overheating.

Tuna and other fish also have oxygen-loving hemoglobin in their blood to facilitate the transfer of oxygen from the gills into the tissues. The deep red color of tuna meat comes from myoglobin, a similar pigment used for oxygen storage in the muscles. Tuna also have relatively large hearts that, unlike most fish, can speed up during periods of fast or long swimming. They do not have swim bladders, however, so they must swim constantly not only to breathe but also to keep from sinking.

Because tuna are such fast and powerful swimmers they are superior predators and easily feed on the less able, smaller fish, squid, and crustaceans of the sea. To sustain their athletic prowess and to grow large, tuna must consume great quantities of food. Like the other bony fishes, tuna use a wide range of senses to locate and identify prey: sight, hearing, smell, and taste. They also have a sensory organ known as a *lateral line* that enables them to detect vibrations and water movement. The lateral line is a network of fluid-filled canals running along a fish's body beneath its skin.

Two other notable characteristics of some of the bony fishes are an internal "antifreeze" system and poison glands. Fish that live in the frigid waters of the high latitudes have special fluids in their bodies that prevent freezing. Other fish have mucus glands that have been modified into special poison-containing structures. These poison glands contain venom that varies in toxicity but can be fatal to other animals. Many of the reef fish, such as the porcupine fish, puffer fish, and trunkfish, and even some of the tunas have naturally toxic flesh. Other fish, such as the inconspicuous and ably camouflaged scorpion fish or the ornately dressed lionfish have spines or stingers that host poison glands. Poison glands and toxicity are a means of warding off predators and signal to others that these fish are not fine cuisine.

Fish must balance the salt concentration of their body's fluids. Seawater is some three times as salty as the bodily fluids and blood of animals. Because water tends to move from areas of low to high salt concentrations, fish are continually leaking water from their bodies into the sea. To compensate for this loss, fish must drink great quantities of water, but in doing so they also ingest more salt, which is then absorbed in the intestines and ultimately excreted through the gills.*

* Almost all organisms that live in the sea, spend part of their time in the sea, or eat things from the sea have a problem with salt. Seabirds, turtles, and marine iguanas have special salt-secreting glands close to or slightly above their eyes. In most seabirds the gland connects with the nasal duct and drips from the tip of the bill. Pelicans have a pair of specialized grooves in the long, upper beak to channel salty excretions to the tip and prevent it from being swallowed (Waller, 1996). Sea turtles often appear weepy as salt is excreted in oozing, gooey tears. True marine iguanas, found only in the Galapagos Islands, feed on algae in the sea, so they too must rid themselves of excess salt. In marine iguanas, the salt gland opens into the nose and a salty goo is sneezed out through the nostrils. Large groups of marine iguanas that live along the dark rocky shores of some of the Galapagos Islands sit around during the day, soaking up the warmth of the sun's rays, and constantly sneeze sprays of salt to relieve themselves of the sea's natural flavoring (often misinterpreted as spitting). Sea snakes have an excretion gland in their mouths, and spit salt out on the tips of their tongues. Plants that live along the sea's edge must also deal with excess salt. The red mangrove, a common shrub along tropical shores with a characteristic tangle of tall orange-brown roots, is a salt excluder. Salt is prevented from entering the plant by special pores on its roots through which seawater is drawn upward. In contrast, both the white and black mangroves are salt excreters, and their leaves become crusty with briney secretions.

Reproduction in tuna, as with many of the bony fishes, is accomplished through the production of lots of small eggs. The female tuna lays millions of small eggs, about a millimeter in diameter, slightly below the water's surface. Sperm released into the water by the male tuna fertilizes many of the eggs. After hatching, the larvae grow rapidly, but predation by others, including some of their own, is high. Many organisms in the ocean produce large numbers of small, poorly developed eggs. The evolutionary strategy is simple: with so many eggs, a few are likely to survive. Another reproductive strategy in the sea is to release only a few, but better developed, less vulnerable eggs or young. Few of the bony fishes brood their eggs or release only a small number of well-developed young; the exceptions include seahorses, pipefish, and surfperch (Waller, 1996). Some of the fish in the sea are hermaphrodites, each fish carrying both the male and female sex organs. And strangely enough, some fish species can initially be either male or female, but later change sex if needed.

While it is sometimes all but impossible to distinguish between male and female fish, in certain instances the differences are exceptionally clear. The male may be more brightly colored than the female, possibly for the purpose of courtship, and the female may become overtly rotund when laden with eggs. In the pink salmon, breeding males develop a hooked jaw and humped back. In one of the most bizarre cases, the male of the deep-sea anglerfish is much smaller than the female and lives parasitically, attached to the head of the female. Most of the bony fishes mature between 2 and 14 years; it takes a bluefin tuna between 3 and 5 years to enter adulthood.

Tuna, along with many of the other swimmers of the sea, return each year to spawn in a specific region of the ocean. From what little is known about tuna behavior, they appear to spawn and spend their early lives in shallow, warm areas and then, as adults, make great migrations or temporary forays into colder and deeper water. It is only within the last few decades, particularly with the advent of sophisticated tagging and satellite techniques, that we have begun to actually track the migrations of the sea's nekton, tuna included. Many seem to migrate in a sort of reproductive equivalent to circuit training in a gym. In one area of the sea, spawning takes place. At another site, the nursery grounds, the young begin to grow and take on more of an adult form. When sufficiently grown, the young move again to the feeding grounds. Once sexually mature, the adults then return to the spawning grounds to breed, and the circle of life begins once more.

Migrations of the ocean's denizens, from fish to mammals, may take place seasonally, over long distances, across fresh and saltwater boundaries, or over a variety of water depths. Many whales migrate seasonally from their winter breeding grounds in the tropics to their summer feeding areas in the temperate zone. Each year, female sea turtles make extremely lengthy excursions back to the same beach to lay their eggs, and in a herculean and near-miraculous effort, salmon swim upriver to spawn in the flowing freshwater streams of their birth. The American and European eels migrate in the opposite direction to spawn, from freshwater rivers and streams back into the salty waters of the Sargasso Sea. After birth, the eel larvae join the ranks of the plankton and are whisked away by the ocean's flow. How they develop and find their way back to freshwater remains another of the sea's great mysteries.* In fact, we know very little about how almost any marine organism skillfully navigates the ocean's unmarked highways.

In some species, such as salmon and sea trout, it is thought that a sort of chemical imprint, or "scent," is imparted to the fish after hatching and serves as a beacon for the returning adult. Fish have a small olfactory

* For a wonderful account of the eel migration, see Rachel Carson's book *Under the Sea-Wind.*

organ, or "nose," to sense chemical traces in the water. Sharks have an exceptionally sensitive nose, particularly for blood (see page 225). To lay their eggs, female sea turtles find their way back to narrow strips of sandy land that are sometimes only small specks in a wide sea. Some suspect that turtles sense light from the stars above and perform a sort of celestial navigation. Tuna and other fish may follow changing patterns in water temperatures or specific water mass signatures. Another theory suggests that some fish and marine mammals may have magnetite-laden bacteria in their tissues that allow them to sense magnetic changes and cruise through the oceans guided by the planet's magnetic field. Marine mammal strandings have been correlated with fluctuations in the magnetic field and areas of mixed magnetic signals, supporting the idea that they use Earth's magnetic field for navigational purposes.

Tuna can live to be more than 30 years old, reaching a length of some 3 meters, and a weight of more than 1400 pounds. Just as the age of a tree can be determined by counting rings in a cross section of its trunk, so can a fish's age be resolved by counting the annual growth rings in its ear stone, the otolith. In general, the smaller, faster-growing fish of the tropics tend to have shorter lives than the larger, slower-growing species of the colder realms. The deepwater orange roughy seems to have found the fountain of fish youth, with an estimated maximum life span of some 150 years (Waller, 1996).

Many of the fish that live within the shallow sea swim together in thick clouds of fluttering fins and flashing scales. By maintaining a large formation of individuals, schools of fish may efficiently cover long distances, confuse would-be predators, and provide safety within a system having a natural perimeter alarm. Much like drafting in a pack of cyclists, schooling may enable individual fish to swim using less energy than if on their own. Finding mates is also easier among a tight-knit group. Research suggests that there are no leaders, but that the school is a constantly changing, though highly consistent, nonhierarchical formation. Within a school, fish seem to maintain a specific distance between one another. When an intruder enters the school, a graceful parting and reassembling of the ranks occurs; however, if the intruder poses a serious threat, the formation may break completely and then regroup later on, once danger has passed. Sight or acoustic vibrations may be used to maintain the rank and file, keeping a school organized. Schools are usually composed of fish of similar size, and may contain either a mix of sexes or just males or females, depending on the species. There is still much to learn about why and how fish school, though one thing seems clear: living in a highly organized and efficient group is particularly useful in the open ocean where few hiding places exist.

The Twilight Zone

The midwater or twilight region of the sea is darker, colder, and calmer than the shallow zone, and there is more pressure and less food. With increasing depth, the weight of the overlying water creates greater pressure. At the surface there is 1 unit of atmospheric pressure; for about every 10 meters (33 feet) of depth in the sea, an additional unit of atmospheric pressure is added; at 10 meters the pressure is 2 atmospheres, at 20 meters it is 3 atmospheres, and so on. Because gas is easily compressed at high pressure, the swim bladders of deep-living fish are filled with low-density fat or oil instead of gas. Midwater organisms tend to be nearly neutral in buoyancy and their bodies have a reduced muscular and bone structure for a lighter load. Cool water temperatures slow their metabolism, decreasing the amount of food they require to survive. In the sea's twilight zone fish are generally small and the gelatinous creatures are particularly well suited for the environment.

To avoid predation, midwater organisms have evolved clever strategies to take advantage of low light levels. As all ambient light comes from above organisms are most vulnerable to predators looking up. Thus some creatures of the midwater are laterally compressed, making them skinny from side to side. They appear as nothing more than a thin sliver from below and are therefore difficult to spot by underlying eyes. Others, like the squid, use counter-illumination tactics; they may emit light that exactly matches that of the downwelling glimmer. Many of the organisms in this region are red or black. Red light has a long wavelength, so it is absorbed quickly in the upper reaches of the sea and is absent in the deeper realm. Nets towed deep in the sea will often return containing large shrimp colored nearly blood red.

Because there is less food in the sea's twilight region, some organisms migrate upward to feed, while others devise clever ways to either attract food or ensure its capture. Unlike the aggressive attackers of the shallow zone, most of the predators in this zone patiently stalk their prey, lurking in the shadows and waiting for the unwary to come close. Some, like the anglerfish, have glowing, tasty-looking lures that hang in front of their mouths, while others, like the hatchet fish and viper fish, have mouths or teeth that are enormous relative to their body size (Figure 46). Luckily, the large-mouthed, toothy fish of the twilight zone tend to be only centimeters long. The gelatinous zooplankton, with their masses of long, sticky, stinging tentacles, do well in the zone of the patient predator.

The Deep Sea

In the very deepest reaches of the sea, organisms are similar to those in the twilight zone, except their specialized traits are even more exaggerated. At depths below about 1000 meters there is absolutely no light from above, not even a downwelling glimmer, so there is no reason to be laterally compressed or use counter-illumination tactics. Fish and other creatures in the abyss tend to exhibit less bioluminescence than those in the midwater domain. The deep ocean is even better suited for a lifestyle

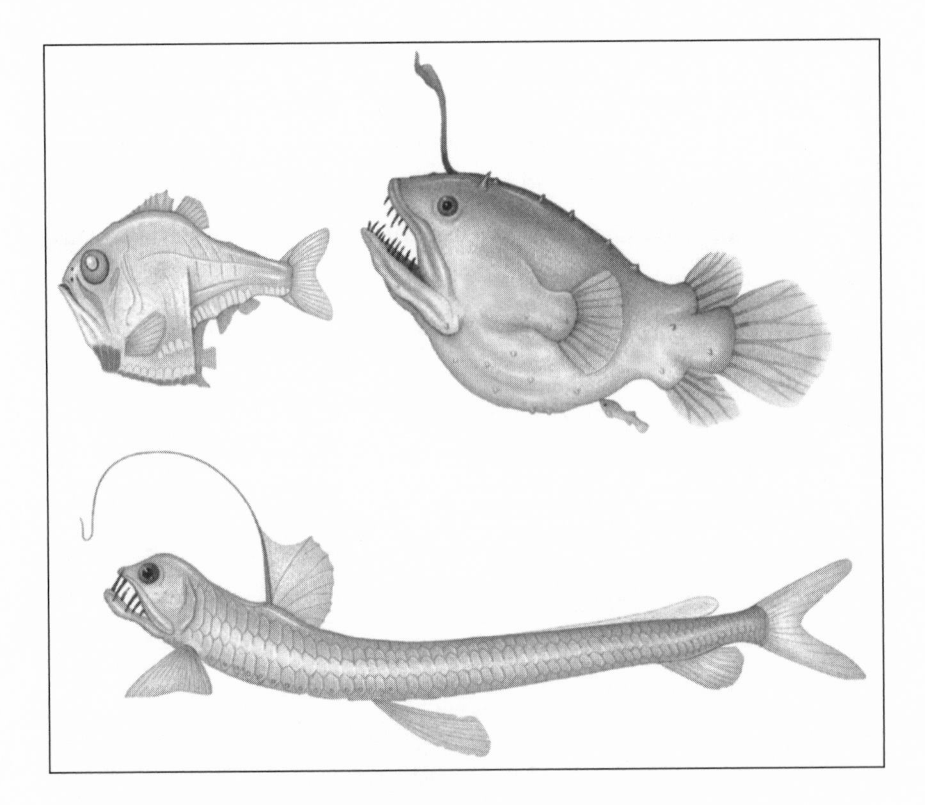

Figure 46 Mid- and deep-water fish.
(Reprinted by permission of John Wiley & Sons, Inc. From Exploring Ocean Science, *Keith Stowe, © 1996, John Wiley & Sons, Inc.)*

of waiting and lunging for prey. With even less food available, organisms must conserve their energy. Creatures are equipped with large, expandable jaws and small, undemanding hearts and kidneys. They loiter in the cold water day in and day out waiting for food to float by, or rain down from above. When large creatures, such as tuna, whales, or dolphins die and sink to the deep seafloor, it is manna from heaven.

Early oceanographers thought that life was absent in the deep sea; today, we are just beginning to discover the strange and wondrous creatures that live there. Recently, the Japan Marine Science and Technology

Center deployed the *Kaiko,* a remotely operated vehicle, into the sea's deepest realm, the Marianas Trench, more than 11,000 meters down. Video footage taken at the seafloor revealed some amazing biological surprises. Scientists found small potato-shaped organisms on the bottom, called xenophores, and swimming sea cucumbers that hovered just over the sediment surface. They also found swimming worms and scavenging eyeless amphipods that functioned by some sort of sensory "smell" organ. The creatures were surprisingly mobile, contradicting the traditional dogma of sedentary lifestyles in the deep sea. Researchers now speculate that these organisms are switch-hitters when it comes to feeding. Most of the time they feed on particulate organic matter that slowly rains down and falls into the sediment; but every once in awhile, when large pieces of food fall down from above, these creatures move quickly, following the chemical scent trail to take advantage of the new food extravaganza.

Squid

Squid are one of the sea's most fascinating creatures. Like their cephalopod cousin, the octopus, the squid are shellfish minus the shell. They are highly adapted to life in the different regions of the sea and are the fastest swimmers of the invertebrates. Most live within the water column, but a few have adapted to life on the seafloor. Adult squid are generally between 20 centimeters and 1 meter in length, though several species, including the giant *Architeuthis,* can grow to enormous lengths. Giant squids with tentacles 15 meters long, and 7-meter-long bodies have been found washed up on beaches or in the stomachs of sperm whales. They are thought to live at depths of about 300 to 600 meters, but no giant squid has ever been caught or photographed alive. In addition to being excellent swimmers, squid can hover, cruise, or dart in any direction. Their maneuverability and lightning quick speed are due to an internal jet propulsion system. Muscle contractions force water into the squid's mantle, the main part of its body, through a pipelike structure called a siphon. The water is then pumped forcefully back out through the siphon, which like a water jet can be aimed in any direction, thus changing the squid from a hovercraft one minute to a jet-propelled missile the next. Stability during its underwater maneuvers is provided by the fluttering translucent fins along its sides. One species of squid is even known to fly out of the water, skyrocketing up to 6 meters high or cruising a horizontal distance of 15 meters before re-entering the sea.

Squid use their excellent sensing abilities and swimming skills to prey mainly on fish and shrimp. To detect a potential meal they use the touch of their tentacles or their highly developed sight. The squid's eye is actually quite similar to that of humans and experiments suggest that it can detect color as well as light intensity. In the sea's twilight zone, some squid have strange asymmetrical eyes. The larger eye is oriented to look up to detect downwelling light and the smaller eye is directed downward to see bioluminescent light (Waller, 1996). Two of the squid's arms are especially long feeding tentacles. They are used to ensnare prey for delivery to the mouth where death comes quickly in a sharp chitinous beak similar to that of a parrot. To confuse prey and escape predators, squid can change color, produce light, and even squirt black or brown ink into the water. Squid are truly among the sea's most intriguing creatures and exhibit complex behavior. In coral reefs, a group of small squid will often approach a diver like a line of marching soldiers. One squid instinctively leads the neatly aligned pack and the rest trail along, in what seems like a highly competitive game of follow-the-leader.

Sharks

Since the movie *Jaws* first introduced the shark as a menacing, powerful, and aggressive eating machine, people have feared and sought to kill these magnificent creatures. In actuality, the behavior and life of the shark are poorly understood; most of what is known comes from those hooked on a line or involved in artificially induced feeding situations. Today, scientists are trying to learn more about sharks through efforts to observe and track them in their natural environment, without influencing their behavior. Results are beginning to reveal that the shark, even the great white, is not a monster that purposely hunts and eats humans, but is a potentially social animal living in the sea, eating and breeding like others, and playing an important role in the overall ocean ecosystem.

Sharks are diverse and widespread creatures of the sea. Some live in the ocean's open, shallow-water realm, while others are better adapted to life in the deeper sea or foraging on the seabed. A few sharks can be found in the cold waters of the poles, while many make their home in the warmth of the tropics. Some sharks make long-distance seasonal migrations and others regularly commute from offshore into coastal waters. Sharks may live mainly solitary lives, but relatively recent work shows that some, like the hammerheads, enjoy the company of others and live in schools. Sharks can be aggressive carnivores or gentle, docile filterers of the sea's plankton. No matter how or where they feed, sharks are at the top of the marine food chain, and like tuna, they play an invaluable role in balancing the ocean's intricate web of life.

Sharks, as well as rays, are among the sea's cartilaginous fish, those having a skeleton of cartilage—a tough, elastic, flexible tissue like that of our own nose and ears. All have relatively long snouts, an underlying mouth, and external gill slits just behind the head. The shape and size of a shark varies depending on where and how it lives. Some of the fast-swimming sharks are tuna-shaped for high-speed cruising (Figure 47), while others are flattened for a more sluggish lifestyle on the bottom. The shark's well-known dorsal fin—like its other stiff, fleshy fins—differs from that of the bony fishes in that it does not have supporting rays and cannot be folded along the body or change shape. The shark's fins

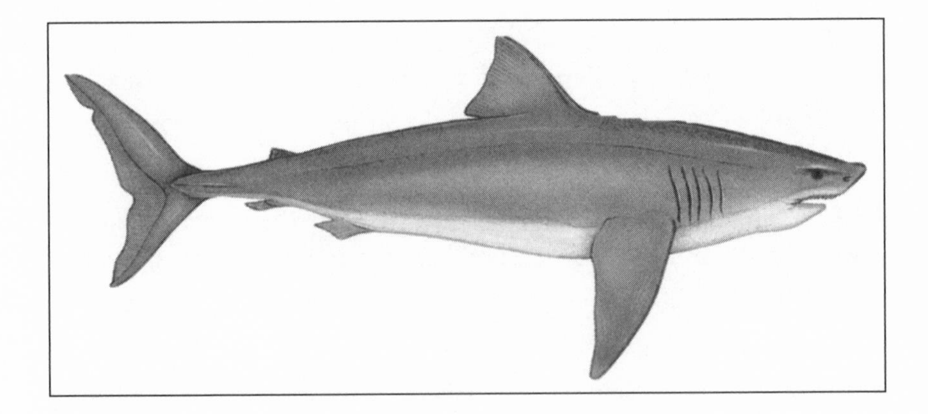

Figure 47 *A fast-cruising, open-ocean white shark.*
(Reprinted by permission of John Wiley & Sons, Inc. From Exploring Ocean Science, *Keith Stowe, © 1996, John Wiley & Sons, Inc.)*

serve many hydrodynamic purposes: they are underwater stabilizers and controls for turning, they provide lift, and they act as brakes. Sharks also use their fins in courtship or defensive posturing. Prior to aggressive behavior, some sharks have been observed to arch their backs and point their pectoral fins downward. The tail or caudal fin of a shark is used to propel its body through the water.

The shape of a shark's tail, much like its body, provides a clue to its watery way of life. It is generally asymmetric, having a larger upper half or lobe. Sharks built for speed have a pointed, tuna-like tail, while more rounded, flexible tails (often with a less developed lower lobe) are characteristic of slower-swimming species. Sharks are propelled through the water by the sideways beating of their tails, which results from a series of muscle contractions that start at the head and progress to the tail. This motion causes the shark's body to undulate gracefully through the water in a serpentine motion. In addition to providing thrust, the shark's tail also helps to steer, much like a ship's rudder.

Sharks and rays do not have swim bladders, so they must swim continuously to avoid sinking. However, their fins, flattened head, and large, buoyant liver help them to maintain their position in the water even at slow swimming speeds. The tough, flexible skin of a shark is made up of an underlying fibrous "carpet weave" of tissue and an overlying lining of aligned, tiny, scalelike teeth. The shark's tough skin helps to reduce drag during swimming and protect it from injury.

On the promotional posters for the movie *Jaws*, a wide-mouthed, toothy shark is shown preparing to chomp down on an unsuspecting swimmer overhead. Though the shark is depicted as a great white, its teeth are drawn like those of a mako (Tricas et al., 1997). Shark teeth come in a wide assortment of shapes and sizes depending on the species of shark and what it feeds on. All sharks have gums that operate like a system of conveyor belts, frequently forming and replacing teeth. Some sharks have serrated triangular teeth for cutting, while others have teeth better suited to piercing, clutching, grinding, or crushing. Bottom-dwelling species have teeth shaped for the crunching and grinding of shellfish and crustaceans. Open-ocean sharks, like the mako, have sharp, triangular teeth better evolved for tearing and cutting.

What makes the shark a lethal predator is not so much its teeth, however, but its extraordinary sensing capabilities. The shark is incredibly well equipped with a wide variety of senses to monitor the surrounding environment and detect prey. For long-distance detection, sharks can use their excellent hearing abilities, sensing sound vibrations at great distances, particularly those of low frequency. The shark also has an extremely acute sense of smell that can detect even minute traces of a substance from very far away. Some estimate that sharks can sense chemicals in the water at concentrations as low as 1 part per million or 10 parts per billion (like a very tiny drop of blood in a large bucket of water). Sharks also have a well-developed lateral line that, like other fish, enables them to detect vibrations created by water movement. Sharks are also equipped with excellent sight. Even from relatively far away, their eyes are capable of distinguishing between light and dark, and can possibly see color. Deep-water sharks and those that hunt at night have eyes especially designed for low light levels. Some of the larger, open-water sharks have an extra eyelid that protects the eye when they are attacking or biting. Even though they cannot see their prey at close range, all sharks are outfitted with an arsenal of other discriminating senses.

Sharks have feelers about their mouths and a network of sensitive nerve endings beneath the surface of their skin, which enhance the sense of touch. They also have an extraordinary close-range electroreceptor system. Pores around the underside of a shark's snout are connected to nerves via a series of jelly-filled tubes. These electrosensors, are sensitive to weak electric fields such as those produced by the heartbeat of potential prey or their movement. Sharks also have taste buds and soft tissue in their mouths and tension-sensitive jaws to help determine the palatability of food. Few shark attacks are fatal to humans because most of the time, after a first nibble and taste, sharks spit humans out. Clearly, we are an unpalatable dish.

While the bony fishes drink lots of water and excrete salt via their gills to maintain the concentration of their body's fluids, sharks and rays concentrate their bodily fluids and excrete salt-rich wastes through the kidneys and rectal glands; thus they do not need to consume vast quantities of water.

Unlike most of the bony fishes that spawn prodigious quantities of eggs, sharks tend to produce only a few internally fertilized eggs. Mating in sharks is not well understood, though biting often appears to be part of the rather rough ritual. Some sharks and rays lay eggs that are encased in a leathery, protective case in which the young feed on an internal yolk sac. The dark leathery egg cases of a ray, sometimes referred to as a "mermaid's purse" or "sea purse," often wash ashore and can be found among the debris on a beach. Other female sharks produce a thin-shelled egg that is retained inside the female until it hatches and a young shark pup is born. Inside the female sand tiger shark, the developing young does not feed on a yolk sac; rather, it prefers sibling cannibalism, eating the other fetuses or unfertilized eggs (Waller, 1996). Another means of reproduction in sharks involves the internal fertilization of eggs and the gestation of young. The embryo develops inside the female shark over a period of months or years, up to 2 years, after which the shark pups are born.

In fish terms, sharks generally grow slowly and live a long time. They seem to reach sexual maturity between ages 6 to 20 and may live anywhere from 7 years up to about 20 to 30 years. The largest shark is the whale shark, which is also the biggest fish in the sea. It can sometimes grow to 18 meters long and weigh as much as two African elephants. However, the whale shark and its large relatives, the basking and megamouth sharks, are not to be feared; they are gentle giants of the sea, swimming through the water with their mouths open, filtering out massive quantities of small plankton. The world's smallest shark is the spined pygmy shark, reaching only about 20 centimeters (8 inches) in length.

The sharks that hunt may do so by fast pursuit or by more stealthy means of ambush. Many sharks are actually foragers that spend their time on or near the bottom searching for shellfish and crustaceans. During the daytime these sharks, such as the common nurse shark, can often be found resting beneath the ledges or crevasses of a reef. They are generally docile creatures, shying away from humans, and foraging mainly at night. But do not be fooled—even the docile nurse shark can be dangerous when provoked. Some sharks are territorial and may react aggressively or curiously to the invasion of their home turf.

Sharks have a large stomach, so that in a single meal they can eat and store large amounts of food. For sharks that live in the open sea, where food may be scarce, this can be quite an asset. Some sharks also have a loosely connected jaw to enable them to get a better and bigger bite when feeding. Because sharks may not take small bites and chew their food before swallowing, they sometimes shake their heads back and forth violently to tear off or rip apart large chunks of meat.

One of the oddest-looking sharks is the hammerhead, with its large eyes perched wide apart on the sides of a flat, broad, rectangular-shaped snout (Figure 48). The flattened snout is thought to serve several functions, providing lift during swimming and increasing the area for senses used in feeding: sight, touch, smell, lateral line, and electroreceptors. The hammerheads tend to be fast, skilled predators, able to capture large, elusive prey. The great hammerhead shark seems particularly fond of snacking on rays and uses the side of its snout to pin the winged creatures down on the seafloor. It then turns its head to the side and neatly chomps off a chunk of the ray's wing. Incapacitated, the ray is at the mercy of the shark, as it consumes its dewinged meal (Tricas et al., 1997).

Another interesting, even more bizarre, feeding strategy is that of the cookie-cutter shark. Whales, dolphins, and large fish are sometimes

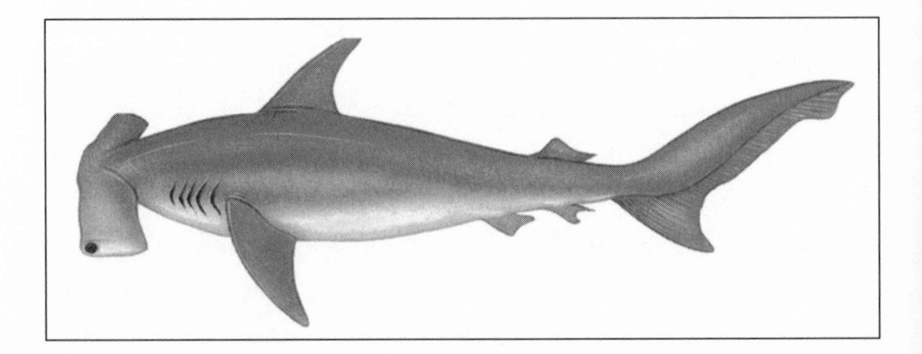

Figure 48 *The hammerhead shark.*
(Reprinted by permission of John Wiley & Sons, Inc. From Exploring Ocean Science, *Keith Stowe, © 1996, John Wiley & Sons, Inc.)*

found with circular plugs of flesh missing—like a hole left in pastry dough from a circular cookie cutter. Underwater cameras and traps have revealed the perpertrator: a small shark with sucker-like lips, sharp teeth, and a strange body that can rotate around its jaws. The cookie-cutter shark feeds by attaching itself to a larger animal, swiveling its lips and jaw around, and expertly removing an oval-shaped plug of flesh from its scarred but living victim.

The story of the rare megamouth shark is also an intriguing one. It is a large, plankton-scooping shark, named from the Greek for "giant yawner of the deep sea." The megamouth was first observed in 1976 and is believed to be a deep-water species. Since one was first identified off the island of Oahu, Hawaii there have been approximately 10 more sightings in Japan, California, Brazil, Senegal, and the Philippines. Why were megamouth never seen before? Are deep-water conditions in the sea changing, prompting this reclusive shark to venture more often into shallower waters, or are the megamouth just one among many unknown creatures that remain hidden within the sea's mysterious deep waters?

Scientists today are studying sharks, not just as specimens hauled into a boat, but as living creatures within their own wet world. Researchers are using ultrasonic transmitters, tagging, and satellite-tracking techniques to discover where these majestic creatures live, eat, and breed in the sea. We are just beginning to learn about their physiology and incredible resistance to disease, and the role they play in keeping the ocean healthy.

However, much like tuna and many other fish, shark populations are being rapidly decimated by overfishing. Sharks are especially vulnerable because they reproduce in small numbers and take many years to grow to maturity.

Sharks are not the evil eating machines once believed. There are few shark attacks and most occur because humans are mistaken for food or sharks have been provoked in some way. Some dive operators increase business by routinely feeding sharks amidst divers. This practice may be teaching the sharks a dangerous lesson: humans equal food. This could potentially lead to tragedy if sharks mistakenly approach unknowing divers or swimmers expecting food.

Like the sharks, their relatives the rays are unjustly feared by many (Figure 49). Rays are physically similar to sharks except that their bodies are essentially one large set of pectoral fins. They swim by flapping their fins, much like the beating of a bird's wings. For the most part, rays are docile creatures that fly gracefully through the sea, feeding on small fish, crustaceans, and shellfish. Many spend their time lying on the seafloor, while others wander the open ocean. Rays have relatively large brains, excellent sensing capabilities, and they exhibit complex behaviors and social interactions. Almost all injuries from rays happen when people accidentally step on a ray's barbed tail. Doing the stingray shuffle (dragging one's feet on the sandy bottom) will typically scare off a shy ray resting harmlessly on the seabed and prevent accidental injury.

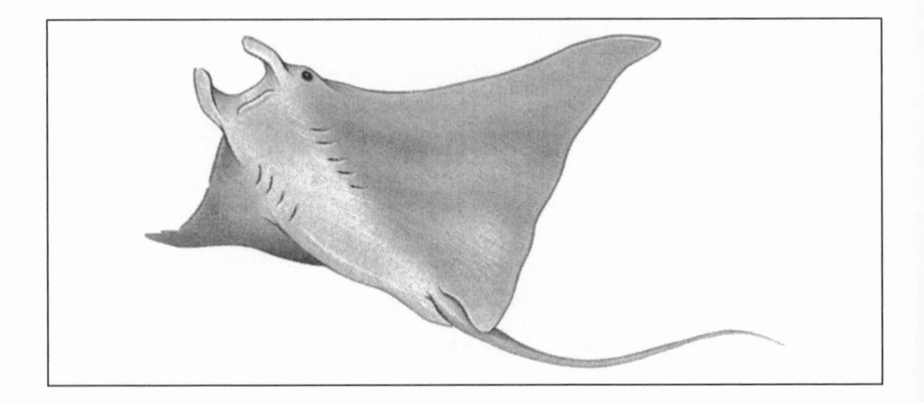

Figure 49 The manta ray.
(Reprinted by permission of John Wiley & Sons, Inc. From Exploring Ocean Science, *Keith Stowe, © 1996, John Wiley & Sons, Inc.)*

Marine Mammals

Marine mammals can be cute, playful, and intelligent, and they are also large enough to be easily observed, thus they have become the ocean's charismatic megafauna and have become the ambassadors for the marine world. Yet in the big picture, marine mammals are no more important than creatures less well known or well liked, such as the gelatinous zooplankton or the tiny copepods. And like other wild animals, marine mammals can at times be dangerously aggressive and quite unpredictable.

The marine mammals include whales, dolphins, sea otters, manatees, dugongs, polar bears, seals, walrus, and sea lions (Color Plate 16). Like other mammals, they are warm-blooded, breathe air, and give birth to live young that nurse on their mother's milk. To stay warm in the cold sea, some marine mammals have fur, while others are encased in a thick envelope of insulating blubber. Blubber or air trapped within the fur also provides buoyancy. Whereas fish swim by flexing their tails from side to side, marine mammals propel themselves by moving their tails up and down. Most of the marine mammals are carnivores, feeding on the sea's smorgasbord of fish, crustaceans, shellfish, echinoderms, and, sometimes, each other. Sea cows, manatees, and dugongs, are the ocean's only herbivorous mammals. They feed on underwater vegetation, such as seagrass and water hyacinths. These creatures are gentle, slow-moving mammals with wide bodies, flat paddle-like tails, and hairy muzzles. They are found only in the warm, shallow waters of the world.

Surprisingly, we still know little about the behavior and lives of the marine mammals, particularly those that spend a good portion of their time in the ocean's deep, darker waters. Much of our understanding of these creatures comes from animals that wash ashore dead or dying, and those that are raised in captivity. Hunting has greatly reduced their numbers, making observation and study even more difficult. Protective legislation and a new respect for the sea and its inhabitants has allowed some populations to increase, but undoubtedly the number of marine mammals in the sea today is nowhere near what is was before humankind's intervention. However, once again, modern ingenuity and technology is beginning to tell us more about where and how marine mammals feed, breed, and live in the sea.

The cetaceans—whales, dolphins, and porpoises—are some of the ocean's most popular and agile creatures. They jump, twirl, somersault, dive, slap their tails, wave their flippers, and often seem to take pleasure in riding along with a ship at its bow. Cetaceans are also curious, caring, sometimes aggressive, and they exhibit complex, though recognizable, social interactions. For the most part, our knowledge of cetaceans comes from observations at the surface. New tagging, tracking, and video techniques to study them at depth are just beginning to produce results.

The main difference between whales, dolphins, and porpoises is, in theory, simply size: whales are big, dolphins medium-sized, and porpoises smaller. However, some whales—the killer and pilot whales, for instance—are dolphin-sized and some dolphins are porpoise-sized. And undoubtedly, those who go to a restaurant and see "dolphin" on the menu think of Flipper, when it is really a dolphinfish, also called *mahi mahi*. To avoid confusion, cetaceans are often divided into toothed and baleen whales. The toothed whales, as their name suggests, possess teeth and they have only one blowhole. The baleen whales have two blowholes, and their teeth have been replaced by large, furry plates that hang down from the gums and are used to filter or sieve food from the sea. The baleen whales include some of the most popular and recognizable of the whales: the humpback, blue, gray, right, and bowhead whales. All of the other cetaceans—including the sperm whale, orcas, beluga, and narwhal, and both ocean and river dolphins—fall into the toothed category.

Toothed whales and dolphins have a greater range and diversity than their baleen cousins, primarily because of feeding preferences. The toothed cetaceans hunt and feed on a variety of relatively large, one-course meals. Many are not picky eaters and will dine on fish, squid, or shrimp. Some may also feed on bottom-dwelling creatures or those that live in reefs. The jaws and teeth of the toothed whales vary depending on their dining habits. Some are fast grabbers and have long, narrow jaws for rapid snatching action. Others have short, broad jaws designed for power rather than speed. Most have conical-shaped teeth for grasping, but some, like the orca, can bite and tear the flesh of their prey. Dolphins grab a fish and swallow it whole, making sure it goes down head first, with the grain of the scales—otherwise it could be a choking, scratchy meal. Sperm whales are the largest of the toothed whales; they have a big, square head and face, with peglike teeth on a narrow lower jaw. In the sea's dark domain, sperm whales are thought to detect prey via echolocation and feed using a sort of slurping suction action. The remains of giant squids have been found in the stomachs of sperm whales and may be one of their favorite deep-sea dishes.

The sperm whales and other cetaceans have physiologically evolved to become masterful deep-sea divers. Male sperm whales have been tracked to depths of more than 2500 meters in dives lasting longer than 2 hours (Carwadine et al., 1998). Upon diving, cetaceans fill their lungs with air and close their blowholes. Their lungs are then collapsed as oxygen is efficiently transferred to the blood and muscles, which are rich in hemoglobin and myoglobin, respectively. At the same time, their heart rate is reduced and blood flow is restricted mainly to the heart and brain. A countercurrent heat exchange system similar to the tuna's is present in the fins and skin and helps to reduce heat loss during diving.

Many of the toothed whales can echolocate with incredible precision. This may be especially important to those species that spend time in the sea's murky or deeper waters. The human equivalent of cetacean echolocation is sonar (sound navigation and ranging). If we could unlock the means by which dolphins produce detailed sound-derived images, our own sonar systems could be vastly improved. However, humans have yet to create a sonar system (echo sounder) as precise and effective as that of the whales, dolphins, and porpoises. Essentially, the cetaceans emit high-frequency sound waves, which sound like a series of closely spaced clicks. The clicks seem to be produced from the soft tissues of the nasal passage, and the melon—a large, fatty internal structure between the skull and blowhole—helps to transmit and project the sound forward as a narrow detection beam. After reflecting off objects, the returning sound signal is picked up by the animal in the sides of its head and jaw and transmitted to the ear. It is unclear how the brain then process the returning sound or echo, filtering out all the other background noise, and creating a detailed "acoustic" picture of the environment. Toothed whales and dolphins also appear to use sound in communication, which may help in cooperative behaviors as well as social interactions. Cetaceans have been observed to work collaboratively to herd prey or undertake coordinated attacks. And groups, known as pods, may remain together for long periods of time, and cluster protectively around wounded members.

In contrast to the toothed whales, baleen whales feed by skimming or gulping huge mouthfuls of water, forcing the water out with their tongues, and then swallowing the remaining food material. The furry plates of baleen that hang down from their upper gums are used as sieves. They have no teeth and eat a wide variety of small marine creatures, including krill, copepods, amphipods, small fish, squid, and pteropods. Though most of the baleen whales feed near the surface, the gray whale is a bottom feeder. It stirs up the mud and strains out or sucks up the small creatures that live in the sediment. On the other hand, the right whale is a skimmer, swimming slowly, mouth open, along the sea surface, filtering food particles as it goes. The humpback whale is a gulper. It feeds by taking short, dramatic gulps. Humpbacks feed as individuals or as a group. Researchers have now observed humpbacks herding their tiny prey into large swarms or using an ingenious technique known as "bubblenet feeding." Diving beneath the surface, the whales exhale huge streams of bubbles that rise toward the surface. Organisms caught within the rush of rising bubbles cluster together tightly, as if caught in a fishing net. One or several whales then swim up through the bubblenet, gulping a mouthful of food-laden water. Some suspect that the baleen whales use their sense of smell to locate particularly dense schools of krill (Carwadine et al., 1998).

While both the toothed and baleen whales produce a variety of sounds, it is the baleen whales that have become so well known for the low-frequency moans and wails popularized as whale song. Male humpbacks often sing a complex and moving song complete with rumbles, grunts, squeals, whistles, and wails. It can last for only minutes or as long as half an hour. In any one region, the resident whales sing the same distinctive song. Researchers are not sure why whales sing or what the songs mean, but they appear to be related to mating or social organization. Humpbacks are also the most active breachers, leaping high out of the water and "belly flopping" back in (Color Plate 16a). Breaching may help to remove barnacles or parasites, provide for a better look at the airbound world, send a message, or simply be a form of play. Whales are also known to "spy hop," bobbing up and peaking about, and some slap their flukes or tails, probably as a means of communication, such as a warning or a defensive gesture.

Cetaceans are found around the world in a variety of habitats. Many whales migrate seasonally over long distances. These whales spend the summer months in cold temperate or polar waters gorging themselves on plentiful blooms of plankton. Toward winter, they, particularly the females, move into the warm waters of the tropics to give birth. Food is scarce, but the warm water is welcome to calves born with little blubber.

While there is still much to learn about whales, dolphins, and porpoises, years of dedicated study and fascination are beginning to pay off. Unquestionably, continued observation and tracking (particularly in unobtrusive ways) will unveil more fascinating information about the cetaceans' oceanic life. One aspect of their behavior that continues to remain somewhat mysterious and disturbing is strandings. Often individuals or groups of whales and dolphins become stranded on the shore. Once out of the life-giving and supporting sea, whales lying on a beach are literally crushed beneath their own great weight. Researchers believe that these strandings may have a variety of causes, including injury, infection, disease, toxins, social factors, food supply, or navigational errors. Most mass strandings involve toothed whales, particularly those that live in close-knit groups. Today, we try to return stranded cetaceans to the sea and rehabilitate those that are injured or sick. Maybe one day when we better understand the cause, we will be able to prevent strandings in the first place.

Seals, sea lions, and walrus are furred, carnivorous marine mammals that spend part of their lives in the sea and part on land. Seals feed and play in the water, while they move to land to rest, warm, breed, and molt to rejuvenate their luxurious fur coats. All are expert swimmers, but only the eared seals are relatively agile on land, walking or running on their fore flippers. Seals are typically divided into true seals, eared seals, and walrus. True seals have no external ears and their furry flippers cannot be turned forward, making movement on land difficult and caterpillar-like. Only the hind flippers are used for swimming. The coral seals like the sea lions (Color Plate 16) have external ears, naked soles on their flippers, and the hind pair can be turned forward. In addition, the eared seals use their fore flippers for swimming and waddling on land. Walrus have no external ears; they walk like eared seals but swim like true seals. All of the seals stay warm in the cold sea with a thick layer of blubber and fur.

The fur tends to be thickest in pups that have yet to pile on the fat. It traps warm air next to the body and is kept waterproof by oil secreted from glands linked to the hair follicles. Like the cetaceans, seals have a specialized physiology that allows them to dive and stay below water for extended periods of time. They have a proportionally high volume of blood that is rich in hemoglobin and muscles concentrated with myoglobin, enabling them to transport and store oxygen while swimming and diving. They also collapse their lungs, exhale, slow their heartbeat while diving, and restrict blood flow to the vital organs. Some seals have been recorded diving to depths of more than 300 meters for up to 2 hours (Waller, 1996).

Most seals prefer colder climes, though some live in the cool temperate or tropical seas. The seal diet consists mainly of fish, crustaceans, some seabirds and even other seals, typically the pups. Walrus dine at the seafloor, feeding on crabs, clams, and worms that they ferret out with their tusks and sort with their whiskers. The seals are a noisy, selfish lot, with breath that smells like fish, a loud bark, and behavior that tends to focus on the individual, rather than a group or mate. The pups and females tend to be curious and playful, while the males can be dangerously territorial. The dominant males fiercely protect their territory and the females within their furry harem. The males also tend to be larger, more powerful, and shorter-lived than the females.

Seals are favorite prey for sharks and orcas. These pelagic or open water predators have been observed to swim into surprisingly shallow water along a beach to snatch a seal snack, and orcas have been known to tip over small ice floats to munch on a blubbery, flippered morsel. To breathe air in the ice-covered polar regions, seals use cracks in the ice or create their own holes by continually gnawing at the ice, a precarious occupation in the Arctic, as a polar bear may lay in wait nearby. Another marine mammal, the polar bear, is found exclusively in the Arctic. Polar bears are large, powerful creatures, with an acute sense of smell, thick fur, and a well-insulated body that keeps them warm even in the cold waters of the far north. Polar bears use their webbed paws to dog-paddle in the water and are surprisingly adept swimmers, even for relatively long distances. The white fur of the polar bear provides camouflage on the ice and may help it to hunt its favorite food, the seals. Killer whales may be their only natural predator, but humans are unquestionably the greatest threat to their survival.

One of the most charming of the marine mammals is the sea otter. Known for its thick fur pelt and cute whiskered face, the sea otter lives exclusively along the coastal shores of the North Pacific. In contrast to the seals, the sea otter can spend its entire life in the water, coming ashore only to take shelter from storms. They are skinny creatures with no protective layer of blubber. Warmth and buoyancy are provided by a thick fur pelt, which must be groomed often to add air and restore its oily coat. Sea otters are typically found at the sea surface floating effortlessly on their backs. In this position they can raft up with fellow otters, use their stomachs as dining tables, carry their young, or drape and anchor themselves with a blanket of kelp. They use their tails and hind webbed feet in swimming and diving, and their front paws to adeptly grab and wrestle food off the bottom. Sea urchins, clams, mussels, and abalone make for fine otter dining. Sea otters are obsessed with eating; to stay warm and active they must consume an amount equivalent to one-fourth of their body weight each day. This would be like a human eating 100 hamburgers each and every day! On the surface, otters use their strong teeth or stones to split open or crush shells that contain the sea's tasty morsels. Sea otters are the only marine mammals known to use tools. Some of the larger male otters will also eat fish. The sea otters also have very strong maternal bonds, and pups have been known to scream for hours when separated from their mother. However, not all is cozy among the otters: mating can be a rough sport that includes nose biting, rolling, and an occasional accidental drowning. Sea otters seem to have few natural predators. Eagles are thought to sometimes snatch pups during the nesting season, and orcas may occasionally consume an otter or two.

Marine Reptiles

The marine reptiles include sea turtles, sea snakes, crocodiles, and marine iguanas. Sea turtles are most commonly found in the ocean's warm waters and use their long flippers for swimming. Unlike many of their land-locked or freshwater relatives, sea turtles cannot completely retract their flippers and head into their shells, so they are prone to shark attack. It is not an uncommon to see a sea turtle missing a flipper or two (Color Plate 15c). Because of their weighty shell, sea turtles have a somewhat reduced skeleton to aid buoyancy, and while they are agile creatures in the sea, they are rather inept on land. Still, in order to reproduce, the female must waddle cumbersomely out of the water onto a beach and dig her nest in the sand. There she lays a clutch of soft, leathery eggs before awkwardly returning to the sea, leaving the eggs behind to incubate in the warm sand. The sex of the baby turtles is strongly dependent on the temperature in the nest, and so can vary with depth. Warmer temperatures beget females, and cooler temperatures males. After 2 months, the eggs hatch and a bevy of tiny sea turtles dig their way up to the surface. Once out of the nest, they make a mad dash for the sea, vulnerable to hungry birds, crabs, and other predators. It is believed that the young turtles find their way to the water guided by the light of the moon and stars reflected off the ocean's surface. Where a brightly lit human development abuts a nesting beach, baby turtles can become confused and fatally go in the wrong direction, away from the water.

Sea snakes are found primarily in the Pacific and Indian oceans. They are some of the most venomous reptiles on Earth, but many are extremely docile and have tiny mouths. They feed principally on eels, fish, eggs, and a variety of invertebrate organisms. In some areas, sea snakes are known to be aggressive and will curl around a diver's extremities, while in other regions they are extremely mild-mannered and shy. In Fiji, children can be seen playing with the black-and-white banded sea snakes common to the region.

One of the most infamous of the ocean's reptiles is the saltwater crocodile, found primarily in Indonesian and Australian waters. Though most stay near the shore, they have occasionally been spotted far out to sea. The Pacific saltwater crocodile can grow quite large, reaching a length of 7 meters and a weight of some 2000 pounds. In Australia and Papua New Guinea, saltwater crocodiles are reputed to be smart, patient, and vicious predators, using their powerful tails to knock prey into the water and even overturn small boats. Nature has equipped saltwater crocodiles with a valve in their throats so that they can open their mouths underwater without drowning. Crocodiles are thus able to grab their prey, drag it underwater, drown it, and then swallow it. A small population of reclusive and endangered crocodiles also lives in the brackish waters at the southern edge of the Everglades in south Florida.

Seabirds

Technically, most seabirds are not considered true swimmers because they spend most of their time flying over the open ocean or only swim part-time on the sea's surface (Nybakken, 1993). However, a few of the seabirds are undoubtedly true ocean swimmers, and many could be called temporary visitors to the sea's nektonic world. Seabirds are found throughout the world, extending from pole to pole, but they are most common in the more fertile regions of the sea. Some live on populated shores, others on distant islands, and a few spend most of their time in or over the water. The shape, size, and bill structure of the seabirds varies and often reflects their flying style and feeding habits. Flight may be accomplished principally by flapping (petrel, terns, and cormorants), soaring (pelican), gliding (albatross), or with a multipurpose wing design (frigate). Almost all seabirds have webbed feet; waterproof plumage; and they feed on fish, crustaceans, and other marine invertebrates. To obtain their food from the sea, seabirds use an array of strategies, including dipping and plucking, skimming, hydroplaning, plunge diving, surface snacking, and underwater pursuit. Some seabirds are also well-known pirates of the air, the frigate and gulls being the most infamous and adept at the art. Seabirds take to land to breed and lay their eggs, and many do so in large colonies that allow for important social interactions.

Of the seabirds, the penguins are clearly the best swimmers of the bunch. Having lost the ability to generate flight, their wings have evolved into efficient flippers. Found mainly within the Southern Hemisphere, penguins have blubber as well as a thick pile of modified feathers to keep them warm in the cold sea. They nest and molt on land but must continually return to the sea to stoke their internal fires with fishy fuel. In the water, they are fast and agile, but on land penguins walk upright and waddle awkwardly. For fast downhill travel, penguins will often toboggan on their bellies. Like sea lions, penguins are a noisy bunch, and each species has a trademark call (Boersma, 1999). Some cackle, others trumpet, and one type, appropriately dubbed the jackass penguin, brays like a donkey.

Cormorants are also efficient swimmers, using their feet as paddles and making frequent short, shallow dives in search of food. In the air, however, cormorants desperately flap their wings up and down, appearing as if struggling just to stay airborne. In contrast, the pelicans are graceful soarers that glide on winds high overhead and then dive like missiles into the sea in search of fish. Another of the ocean's bird torpedoes are the boobie birds. In the Galapagos, the blue-footed boobie does a high-stepping, clownish courtship dance on land and then takes to the air in preparation for its bombardment of the sea. However, landings and takeoffs are not a graceful feat, and often a boobie seems about to abort its flight just before becoming airborn.

Only about 3 percent of the world's birds are ocean-going, but they are a fascinating lot and play an important role in nutrient recycling and the overall food web in the sea.

The Bottom Dwellers

A vast array of marine environments and their complex underwater topography allow many organisms to live on, in, or near the seafloor. Along the ocean's rim there are sunny meadows of seagrass, shadowy forests of kelp, and wide sandy beaches. Saltmarshes, mangroves, tidal creeks, coral reefs, and rocky intertidal areas also grace the shoreline. On the floor of the deeper sea, there are vent or seep environments, as well as extensive rocky or muddy habitats. Within each of these settings, those living on the bottom may spend their time in the sediments or be strictly creatures of the surface. Some are mobile—crawling, slithering, or burrowing—and others are fixed in place—sessile creatures that take up permanent residence in one particular spot. Stationary creatures can be encrusting (bryozoans, algae), have roots (seagrass), excavate or bore into rock (sponge, sea urchins, and mollusks), and a few actually cement themselves onto the seabed (oysters). Both physical and biological factors strongly influence the benthos. In coastal environments changing temperature, salinity, and wave energy along with competition predation, and food availability can dictate how and where benthic creatures live and their physical appearance. Space on the bottom is often at a premium, and competition for a spot on the seafloor can be fierce.

For the most part, we know more about the creatures of the bottom than we do about those in the overlying water because many are sedentary or slow and are thus unable to avoid our attempts to collect them.

Bottom grabs, box cores, nets, dredges, scuba gear, submersibles, or remotely operated vehicles can all be used to successfully sample the ocean's plentiful benthos. (As could be expected, we know less about the organisms at deep sites or those living below the sediment surface than the creatures on top of the seafloor and in shallow water.) From these collections we have learned quite a bit about how benthic creatures feed, something about their reproduction, and a little about their behavior. Important information about the ways of life on the ocean's floor has also been gained through observations in the wild and in captivity. But our understanding of benthic environments, particularly deeper ones, is still limited by our restricted access to the sea.

One effective means of spending extended periods of time underwater to study the creatures of the sea, particularly those living at the seafloor, is to live in an undersea laboratory. Such habitats allow scientists to live and work for relatively long periods of time underwater. While aquanauts (someone living underwater for 24 hours or more) eat, sleep, and do some work inside an undersea laboratory, a large portion of their time is typically spent diving with scuba gear. For instance, while living at about 17 meters, scientists can dive, each day, for about 9 hours and go to a depth of 34 meters. This is at least 10 times the amount of time a scuba diver from the surface could spend at similar depths.* While there were numerous undersea laboratories in the past, today the only currently operating underwater habitat is the *Aquarius 2000*, located about 17 meters below the surface in the Florida Keys and operated by the National Undersea Research Program and the University of North Carolina at Wilmington (Color Plate 7). The *Aquarius 2000* and its many predecessors have provided scientists unprecedented

* However, after living beneath the sea for 24 hours or more, all aquanauts must go through some 17 hours of decompression before returning to the surface.

access to the oceans, and in particular to coral reefs. By living among the fish, we learn not only about their world but how we as humans adapt to life underwater.

Lifestyles of the Crawling and Attached

The sea's bottom dwellers are typically categorized by size and feeding method. Essentially there are large (≥ 0.5 millimeter or larger), medium (0.1–0.5 mm), and small (less than 0.1 mm) sized benthos. Large creatures live mainly on top of the sediment surface, while smaller things live within. Bottom fish, octopus, lobster, shrimp, crabs, sea urchins, and starfish are among the bigger benthic organisms. Also within this category are the shellfish, corals, sea anemones, sponge, algae, and some of the worms. Those in the medium-sized range include worms and bottom-dwelling diatoms. The smallest are the microbes, mainly the bacteria. Among the benthos, large and small, there are a wide variety of interesting, ingenious, and sometimes frightening means of feeding.

The simplest method of acquiring food is to make your own through photosynthesis. In the benthic realm, as in the water itself, this is accomplished by plants and algae that live mainly in the sea's shallow, sunlit region. Seagrasses (Figure 50) are the only true land plants that have successfully adapted to life fully submerged within the ocean. They are found throughout the world over a wide range of latitudes in relatively shallow water. Other plants, such as mangroves and saltmarsh grasses, live in the sea but are only partially or periodically underwater. Like meadows of grass on land, undersea fields of seagrass provide shelter for other organisms, are a source of food for grazers, stabilize the underlying sediments, and through the decay of dead plant material add important nutrients to the ocean ecosystem. In fact, seagrass grows so fast (up to 10 millimeters per day; Zieman, 1982), that they are thought to have a major influence on the fertility and health of nearshore marine environments. Seagrass

Figure 50 *Looking into a lush field of seagrass in Florida Bay.*
(E. Prager.)

beds, like mangroves, wetlands, and some estuaries, also provide a nursery ground for the young of many commercially important species. Some creatures live literally on the seagrass, being attached or encrusted onto the blades. Other organisms remain within the shadows of the gently waving canopy. In coral reef environments, lobsters, sea urchins, and fish may leave the safety of the reef at night to forage in nearby seagrass beds. Hordes of black, spiny sea urchins have been observed marching off the reef into the seagrass beds to feed at night, only to return as the light of day approaches.

Algae that live on the ocean's floor are an extremely diverse and colorful group. Some resemble undersea tumbleweeds, rolling and drifting over the seafloor. Others create a plush carpet on the bottom that protects the underlying sediment from erosion but can smother those below. Many algae attach themselves to the bottom by encrusting or creating a short, rootlike holdfast. Some of the attached species have stalks with wide, leafy fronds, while others form short bushes with thin, intertwining branches. In the tropics a great assortment of algae abound. There is *Halimeda,* with its strange flakey plates that create green chains whose links later become sediment. Or *Acetabullaria,* the "mermaid's cup," which has a delicate stalk topped by a small, light green cup with a tiny white ball of calcium carbonate at its center. There are red, green, and brown benthic algae, and many of the encrusting coralline algae are purple or pink. Some algae can grow large in proportion, like the giant kelp common to temperate and cold regions. Kelp have a strong holdfast, a long stalk, and large leafy fronds or blades. The kelp can grow to immense lengths, up to 30 to 40 meters long, and create dense, shadowy forests. Kelp and some of the other larger algae have gas-filled floats to keep them buoyant, thus nearer to the surface and the life-giving sun. Like seagrass meadows, the kelp forest hosts a wide array of organisms that live on, in, or below its undersea canopy. Sea urchins can play an important role in the kelp forest, feeding directly on the algae, keeping its growth in check. In the Pacific Northwest, kelp-munching sea urchins are, in turn, consumed by the constantly hungry sea otter. Consequently, changes in the sea otter population can have a dramatic impact on the kelp beds. Scientists have shown that when otter populations decline, sea urchins proliferate. And if the sea urchin population grows too large, the kelp forests can literally be eaten away.

Conversely, if the population of sea urchins declines, the kelp grows untamed, while the sea otters go hungry.

Diatoms, dinoflagellates, and bacteria are also an important component of the benthic community. Dinoflagellates often live in association with other organisms at the seafloor, such as the large benthic foraminifera. Probably the most well-known and studied example of this symbiosis occurs in corals whose tissues can harbor the dinoflagellate *zooxanthellae*. As an internal and photosynthesizing partner, zooxanthellae help the coral get rid of wastes and build a limestone skeleton. The algae also provides the coral with oxygen and imparts it with a greenish-brown color. In return, the dinoflagellate receives nutrients and protection from its coral host.

Benthic Dining

Within the bottom-dwelling community, those that cannot make their own food must find another way to obtain nourishment. Suspension feeders are organisms that filter organic particles suspended in the ocean's flow. Some do this by passively sieving the water and others actively pump it through their bodies. A sponge may be the most efficient biofiltration system ever, capable of pumping an amount of water equivalent to its own body volume every 10 to 20 seconds and filtering out 99 percent of the particulate material. A scientist once estimated that most of the water overlying a reef could be filtered through the existing sponges in just 2 to 3 days. Consequently, sponges are thought to play an important role in keeping the water over coral reefs clean and clear. Exhibiting brilliant and varying shades of red, orange, brown, white, black, and yellow, the sponges are also a diverse and colorful lot. One particularly beautiful reef sponge forms a delicate, almost iridescent vase of purple. Sponges also take on a wide variety of shapes, forming tubes, baskets, or little more than a thin sheath over the bottom. For defense against the sea's hungry sponge-eaters, some have tiny needles or starlike clusters of silica in their tissues, while others house toxic chemicals. One bright orange or purple reef sponge uses acid, created internally, to literally bore into the coral limestone. Scientists today are studying such sponge compounds for their potential use in medicine. One sponge substance has already shown to have anti-inflammatory qualities, while another could help to reduce rejection during organ transplants. But the sponges are a difficult group to study, species are extremely hard to distinguish, and there are relatively few experts in the field.

Many of the shellfish are also among the sea's filterers. Using their muscular foot, some shellfish burrow down into the sand or mud and then extend a tube, called a siphon, up into the water. Water is "inhaled" through an incoming siphon, filtered for food and oxygen, and then "exhaled" through an outgoing siphon. The depth to which a clam can burrow often depends on the length of its siphon. The common cockle has a short siphon so it must stay near the surface. However, the Tellin clam has an extremely long siphon, so it can dig deep, while still vacuuming up food particles from the water far overhead (Waller, 1996).

The barnacle is another filterer. Though it looks like a shellfish, the barnacle is actually a crustacean, appearing more like its leggy relatives early in life during its larval stage. As an adult, the barnacle attaches itself to a hard surface and secretes a protective shell around its body. Its feathery legs then become feeding appendages, extending out through the top of its shell to scoop up food particles. Barnacles are widely distributed throughout the oceans, and their mobility as larvae allows them to disperse and attach to a wide assortment of hard surfaces, including boat hulls, turtle shells, and even whales. They can also effectively close their shell to avoid drying out or incoming freshwater, and thus live in areas inhospitable to less adaptable marine creatures.

Extending sticky tentacles out into the water is another technique for capturing food particles within the ocean's flow. In coral reef environments the tentacled suspension feeders include a strange assortment of tube-dwelling worms. The Christmas tree worm has brightly colored and spiraling sticky arms, the featherduster looks like a dust mop, and the tentacles of the spaghetti worm look like stringy pasta noddles, as the names implies (Color Plate 14).

Corals can be considered full or part-time members of the suspension-feeding family. At night, or in some cases throughout the day, a coral extends its polyps, rings of short, stubby, stinging tentacles, into the water and captures zooplankton and other organic matter that floats or swims by (Color Plate 13b). Sea anemones do the same, but on a constant basis, capturing food particles with their tentacles and bringing it into the mouth for transport to the gut. Suspension feeders are especially abundant where the water is clear and the bottom sandy. If there is too much sediment in the water it will clog up an organism's feeding apparatus. Nature has, however, provided the corals with their own windshield wiper for the sea's rainfall of sediment. If a light to moderate amount of debris falls on a healthy coral, it secretes mucus that is sloughed off its body along with the sediment. When drilling a coral head for a core sample (see Figure 38), geologists often get slimed by a thick coat of coral mucus. (Such drilling does not usually kill a coral; also, the core hole can be filled with underwater cement so that the coral can grow over the extracted area.)

Many of the sea's creatures stripmine the ocean floor for food. These are the deposit feeders, slurping, sieving, and straining the sediments for small, edible pieces. Organisms like the sea cucumber and sea biscuit (Color Plate 3) plow their way through the sediment, ingesting it, stripping out the organic debris, and then releasing the rest as waste material, often in the form of consolidated pellets. Sea cucumbers may use sticky tentacles to sweep or vacuum up the sediments, or they may simply suck up everything in their path. The sea cucumber is an interesting log-like creature; and comes in a variety of colors, patterns, and textures. Because benthic creatures are typically slow and lack maneuverability or may even be stuck in place, nature has provided them with some ingenious means of deterring would-be predators. Sea urchins have sharp spines, some even with a toxin in their tips; sponges and some worms have bristly spicules or stinging chemicals. But the sea cucumber has a completely different yet extremely effective method of deterence: it eviscerates. When threatened, the sea cucumber shoots its internal organs out through its anus at the attacker. The guts of the sea cucumber are sticky tubules that can cling to a predator like a net, distracting the attacker and making time for the sea cucumber to lumber safely away. Then, like a starfish that breaks off an arm, the sea cucumber regenerates its insides. Deposit feeders, like the sea cucumber, may be picky eaters, choosing specific particles or microbes from the sediments, or they may just eat everything edible. Many worms and some of the crustaceans also stripmine the sediments for food. Sediment feeders are particularly abundant where the seafloor is covered in more muddy, organic-rich material.

Within the benthic realm there are also herbivores, the grazers of undersea algae and plants. Many of these bottom-dwelling vegetarians have specially designed feeding equipment to scrape algae off rocks and corals or to actually grind them up. Snails are excellently equipped grazers, having a tonguelike rotating radula that is covered with razor-sharp teeth. Moving slowly over an algae-covered surface, the snail "licks" the underlying substrate and scrapes off its juicy food. Because the teeth on the radula are continually being worn down, snails produce new, sharp teeth at the radula's back end and then slowly rotate them forward as replacements. To provide lubrication while cutting, grinding, and scraping algae, many snails and their relatives produce slimy mucus. The conch, a large snail once common throughout the tropics, produces enormous quantities of mucus that is extremely slimy in nature. Water and soap cannot remove conch slime; an acidic substance such as vinegar, lemon, or lime juice is needed to do the job. The precise arrangement and shape of a snail's teeth on the radula varies according to its feeding preferences.

Snails can graze for algae in places few others can go. Rocky tidal areas, mangroves, and saltmarshes often teem with small periwinkles, which, unlike most marine animals, can withstand the periodic retreat of the sea. At low tide, the periwinkle simply withdraws into its shell, closes the hatch (a thin, hard, protective plate on its muscular foot), and uses a tight seal of mucus to attach itself to the underlying surface. Left high and dry by the tide, the organism inside thus remains moist. When the ocean's flow returns, the snail cracks the seal, pops the hatch, reemerges, and begins to graze.

Sea urchins can eat both plants and animals, and like the snails, they are equipped with a special tool for scraping and grinding. Beneath its prickly body, the sea urchin houses an organ called Aristotle's lantern, which looks like a flower with its petals all closed in. But in this flower, the overlapping petals are made up of hard, pyramid-shaped, calcareous plates. Muscles run the length of the structure and sharp, pointy teeth lie at its tip. Aristotle's lantern is an extremely effective scraping tool that can be moved over a surface, raised, lowered, or swung side to side. Another notable aspect of the echinoderms, including sea urchins and starfish, are their tube feet: thin, tubular strings with sticky balls on the end, which function as a sort of hydraulic locomotive system. To move, the echinoderms inflate their tube feet with a seawater-like fluid kept inside their bodies. The feet then extend out and forward, and a gooey substance at their tips adheres to the underlying surface. The tube feet are then deflated and retracted, pulling the starfish or other echinoderm forward (or backward since they have no head or tail). Their sticky grip is then released and the process repeated. If you carefully pick up a starfish or sea urchin, you can easily feel the sticky, gripping action of its tube feet.

There are also those among the benthos that prefer meat—the hunters and scavengers of the seafloor. Many of the seafloor's carnivores are picky eaters, but others, particularly the scavengers, eat just about anything, including their own kind and the dead or dying. Benthic organisms use a wide variety of senses, including sight, smell, touch, or chemical detection, to seek out their prey, and they are very creative in how they capture and eat it.

In certain snails the radula is not simply a tool for grazing, but a dangerous weapon that can saw, drill, or pry open the shell of a mussel, clam, or oyster. The radula of the necklace shell snail is even tipped with sulphuric acid to accelerate the process of boring through the shell of its prey (Waller, 1996). Some snails also inject a deadly toxin into their victim. The most effective weaponry is that of the cone shell snail. Its radula can be thrust like a harpoon into its victim, and the tip is laced with a deadly neurotoxin. In the Pacific, particularly venomous species of the cone shell have caused a few human fatalities.

From our perspective, the starfish seems a benign and unassuming creature, lying beneath the waves and moving ever so slowly. Yet for many that live on the seafloor, the starfish is one to be wary of, a hungry predator feeding on shellfish, corals, and sometimes other starfish. In one of the more interesting but messy means of undersea dining, some of the starfish can actually force their stomachs out of their bodies, engulf their prey, digest it, and then retract the satiated stomach back in. Their sucker-like feet can be used to pry apart oysters, clams, or other shellfish, allowing the mobile stomach to invade the helpless mollusk and digest its tasty contents. On the Great Barrier Reef in Australia, one of the most voracious carnivores is the Crown-of-Thorns starfish. It chooses a tasty-looking coral, spreads its stomach out on top of it, and creates a soup from its digestive enzymes and the underlying coral tissue. When it retracts its stomach, nothing is left but a lifeless, white skeleton below. In recent years there have been several major outbreaks of the Crown-of-Thorns starfish in Australia that have caused widespread coral damage on the Great Barrier Reef. It is unclear what caused the outbreaks, but many suspect that land runoff high in nutrients created phytoplankton blooms that in turn provided an overabundance of food for starfish larvae. Others believe that overfishing removed the starfish's main predators, allowing the population to periodically bloom.

The octopus is one of the most avid and skilled hunters of the seafloor. It is a smart and persistent predator that may search out food or lie in wait camouflaged within a rocky bottom. Once caught, its victims have little chance against its eight, sucker-covered tentacles and toxin-laced jaws.

A few of the crustaceans, particularly some of the smaller crabs, are filter feeders, but most, including lobster and shrimp, are hunters and scavengers. Some of the shrimp are the most accomplished and well-equipped hunters. The mantis shrimp is particularly fond of preying on fish, crabs, and shellfish. It has highly evolved eyes that rotate independent of one another and can detect color, light, and very faint water movement. Once prey is sighted, the mantis uses its leggy appendages to either spear or crush its intended meal. Another shrimp, equipped with its own stun gun, uses sound to incapacitate its prey. These snapping shrimp are common in the craggy interstices of a reef. They snap their legs near their intended prey with such force that it creates an incapacitating shock wave. Hydrophones (underwater microphones) placed in the water overlying a reef clearly pick up the telltale snapping and clicking of these adroit, but small predators. One of the more gruesome predators is the painted dancing shrimp. It is a secretive and cunning creature that lives in pairs within the reef. Nocturnal hunters, the shrimp seek out a starfish, capture it, and drag it back to their lair. The captive starfish is then flipped upside down to prevent escape and over a period of days, sliced up, arm by arm. Thus, the still-living starfish provides the devious shrimp with a lasting larder of ultra-fresh food (Waller, 1996). On the whole, most crustaceans are scavengers, eating the maimed, dying, or dead of the seafloor and often resorting to cannibalism. Luckily for humans, the sea's treacherous benthic world is one dominated by the diminutive.

When they are not eating, the benthos spend much of their time trying to avoid being eaten. Some can simply swim away from their would-be attackers, while others must use alternative defense tactics. Even the gangly lobster can swim, if need be. The lobster usually avoids the sea's hungry masses by staying hidden within the cracks and crevasses of the seafloor. However, if caught out in the open it can swim backwards away in a flash, using a rapid snapping action of its tail. Many others also seek refuge in the seafloor's hiding places or, if they are of the shelled variety, within their own protective armor. Some of the bottom dwellers also burrow away from trouble, while others are equipped with spines, toxins, or specialized defense mechanisms, like the eviscerating tactics of the sea cucumber.

Another means of protection commonly found at the ocean's floor relies on a partnership for security. Probably the most well-known and photographed example of this protective symbiosis is that of the cute Pacific clownfish and its tentacled host, the sea anemone. The small, striped clownfish live in pairs within the protection of an anemone's stinging arms, and in return, they ward off the anemone's potential enemies. The clownfish could in fact be said to sleep with the enemy, or more appropriately,the anemone. Further study of the organisms of the seafloor is revealing an increasing number of highly organized and complex interactions between species. Among some coral species there is a hierarchical order of dominance, sort of a king-of-the-hill arrangement. While some species can live in peaceful undersea bliss as neighbors, others spark coral warfare when situated next to each other. When two unfriendly foes abut tissues, battle erupts, with one coral assaulting the other using stringy mesentarial filaments. In the coral hierarchy certain species always win the war against specific neighbors, and thus expand their territory at the adjacent coral's expense.

Reproduction among benthic creatures occurs through a variety of asexual or sexual means. The sea squirt, a relatively small, simple bud-shaped creature, lives attached to the seafloor typically in clusters. Individual sea squirts reproduce by simple budding, essentially self-cloning. Some of the corals also do this; others regenerate through fragmentation. When a piece of a branching coral breaks off, if conditions are favorable, the fragment can simply regenerate and begin to grow as a new coral colony. Corals also reproduce by spawning, releasing eggs and sperm into the water. Within a reef, corals may all spawn around the same time each year, almost to the day. In Florida, the Caribbean, and the Gulf of Mexico, this momentous occasion usually takes place in August, about one week after the full moon. During spawning, the water becomes cloudy as a flurry of tiny, round coral eggs and stringy sperm float to the surface (Color Plate 13c). Mass coral spawning is a spectacular undersea nighttime wonder few are privileged to behold. Worms and other small creatures are drawn by the abundance of edible coral eggs and sperm in the water, and a feeding frenzy ensues. If the coral eggs survive and are fertilized, they develop into young planktonic larvae. Many of the other benthos, including crustaceans, shellfish, echinoderms, worms, and sponge also produce planktonic larvae, which may spend hours, days, or even months floating within the sea before settling to the bottom. How, where, and why larvae choose a place to settle on the seafloor is a topic of great interest to researchers. Some appear attracted by a chemical trace, possibly from individuals of the same species or another related species. Environmental conditions such as temperature, currents, salinity, light, and bottom type also seem to play a role in influencing when and where a larva makes its descent to the bottom, and grows into adulthood.

A wide variety of physical and biological factors influence how benthic creatures live and relate to one another. The ecology and biology of the sea's bottom dwellers is too lengthy a topic to describe here, but one specific seafloor environment deserves special mention: the deep-sea vents.

Deep-Sea Vent and Seep Communities

The discovery of an entirely new and unexpected ecosystem thriving at active hydrothermal vents in the deep sea was unquestionably one of the most spectacular events in biological oceanography in the 20th century. In 1977, when geologists first described the strange abundance of bizarre life-forms at deep-sea vents, biologists were listening in at the surface. At first, their reaction was sheer disbelief—what would geologists know! But as observations poured in and images were unveiled, the biologists were not only convinced, they were completely amazed. No one had ever imagined that such abundance and odd forms of life could exist deep within the ocean in an environment devoid of sunlight and awash with potentially toxic compounds. Now it seems that these gardens of deep-sea life are quite common, having been found throughout the world's oceans at hot and warm hydrothermal vents, as well as at cold gas and oil seeps. Our limited sampling and access into the deep sea had for so long kept an entire ecosystem hidden from view. More than 300 new marine species have now been found, including new types of clams, tube worms, shrimp, crabs, mussels, siphonophores, and bacteria. Unlike almost all other ecosystems on Earth, deep-sea vent communities do not derive their energy from the sun through photosynthesis, but are based on chemosynthesis, using only chemical compounds to grow and flourish.

The most widely recognized vent or seep creatures are undoubtedly the large red tube worms that grow in luxurious thickets at many sites with one aptly named, the Rose Garden. The tube worms can reach more than 2 meters in length, and their high hemoglobin content gives them their blood-red color. Studies of the worm's physiology reveal that they have no digestive tract and appear to obtain nutrition from bacteria living within their bodies. Vent clams, in excess of 25 centimeters in length, also harbor the bacteria, as do mussels similarly found clustered around active seeps (Color Plate 1). At vent and seep sites, bacteria are found in great quantities, creating clouds in the water and dense mats over the seafloor. In a sense, the bacteria are the deep-sea equivalent of phytoplankton in the ocean's shallow, sun-driven food web. Bacteria are the primary producers at vents, using hydrogen sulfide to create energy for the synthesis of organic material. The larger organisms, such as crabs and shrimp, feed on the bacteria, or like the tube worms, use them internally to grow. The tube worm's blood contains a special protein that binds to the hydrogen sulfide in vent water and transports it to the bacteria living in the worm's body. The bacteria then use the blood-bound sulfide along with carbon dioxide in the water to create organic carbon within the worm. How this is done is not exactly known. Tube worms and clams in cold gas or oil seeps appear to have a similar physiology, but their bacteria use methane instead of hydrogen sulfide for chemosynthesis. Bacteria within these deep-sea gardens also provide food for fish, snails, masses of stringy spaghetti worms, and a white or pink puffball creature, now identified as a siphonophore.

Like many of the other benthic environments, deep-sea vent communities also show an ecologic zonation, wherein the distribution of species is a function of distance from the active plume. Temperature may be the controlling factor, with those unable to tolerate high temperatures living further away from the escaping, superheated water. In shallow water environments, variables such as wave energy, tide line, or grazers can control the distribution of species. Dead tube worms and clams found at inactive vent sites suggest that these oases of abundance in the deep-sea may be fairly short-lived or ephemeral features. Research also shows that variability in the flow at a vent can strongly impact the composition and health of the surrounding animal population.

All live vent and seep communities are centered around active plumes. Consequently, hundreds or thousands of kilometers may separate one living community from another. One of the more intriguing questions about vent organisms is how they initially find and colonize active vent sites. Many of the major vent animals live attached to the seafloor but have planktonic larvae, so undoubtedly it is the larvae that colonize newly active regions. But how do they find such sites in the vast, deep reaches of the sea? Talk about finding a needle in a haystack! The larvae must be able to sense heat or a chemical signal that leads them to a vent or seep site once it becomes active. Scientists studying bacteria and other organisms that thrive on dead whale carcases in the deep sea, now speculate that such sites could provide the stepping-stones for the larvae of vent animals (Dybas, 1999).

The bacteria that live in such hot, hostile, and extreme conditions are often called extremophiles or thermophiles. In 1996, analysis of the DNA structure of one bacteria species revealed that its genes were completely unrelated to previously known genes (IWCO, 1998). Consequently, researchers believe that some of the bacteria may not actually be bacteria but a new form of microbial life, now named *Archea,* for "ancient ones." It is also not clear how far down into Earth such bacteria, or *Archea,* can exist, or how abundant and diverse they really are. Some bacteria have been brought to the surface and cultured. Because they use hydrocarbon or sulfide products to grow, it is hoped that such bacteria may provide an environmentally friendly means to clean up oil spills and other toxic wastes.

The abundance, diversity, and growth of life at active vent and seep sites are greater than ever expected in the deep sea. Life in the ocean's deepest reaches appears not to be limited by the extremes of pressure and temperature, as was once believed, but simply by the availability of food. And as mentioned earlier, some scientists believe that life on Earth may actually have originated in vent-type conditions. Today, we continue to study deep-sea vent biology and try to answer the many fascinating questions posed by these recently discovered and magnificent marine ecosystems.

Limiting Life in the Sea

Within the sea, be it in the water or on the bottom, all organisms are heavily influenced by their surroundings. Creatures must deal with physical and chemical factors such as temperature, salinity, light, sediment, oxygen, pressure, currents, sea level, and wave energy. They must also cope with biological influences such as food or nutrient availability, competition, and predation; and not only do these factors vary from place to place and region to region, they change with time. In fact, it is often the extent of change or variability within a marine environment that sets limits to life's prosperity in the sea.

When the ocean is said to be highly productive or fertile, it generally means that it is rich with life. Such prosperity typically begins at the base of the sea's food web and spreads upward, from the algae and bacteria to the sea's fish and large mammals. In the shallow sea, the growth of the lowest-end members—the plants, algae, and photosynthesizing bacteria—is controlled principally by two things: the availability of light and nutrients. At the very surface of the sea, the extremely strong sunlight is too intense for many of the phytoplankton. At the same time, they cannot go too deep because light levels become too low. Even in the clearest of clear water (typically the tropics), 90 percent of all red light is gone by about 10 meters, and by a depth of 150 meters only 1 percent of blue light remains. So plants and algae are thus compelled to live near the surface, but not too close. However the surface zone tends to have few of the nutrients they need for growth.

Nutrients are simply chemical (inorganic) substances necessary to maintain life in the sea. Two such chemical materials needed for photosynthesis are nitrogen and phosphorus, and thus their availability, particularly at the surface, is believed to be one of the most important factors limiting the ocean's productivity. Other elements, such as iron and silica are also essential in the production of organic matter. Diatoms and radiolarians require silica to make up their shells, and recent research has shown that, particularly in the Southern Ocean and other nutrient-rich regions, the amount of iron available can limit growth.

In the ocean, dissolved nitrogen and phosphorous originate principally from the breakdown or decay of organic matter (and some comes from land runoff). At the sea surface, where most of the marine plants and algae reside, as soon as nutrients become available they are quickly consumed. As a result, concentrations of nutrients near the surface tend to be very low. In the deeper sea, where there is more decaying organic matter and few, if any, photosynthesizing creatures, nutrient levels are higher. Mixing by wind or upwelling can bring nutrients up from below to the surface and create a bonanza of marine life. However, in many regions nutrients become trapped in relatively deep waters by a density barrier created by changes in temperature with depth.

Most of the ocean is warm on top and cold on the bottom. Between these layers lies a region of change, where the less dense, warm waters of the shallow give way to the more dense, cold waters of the deep. It is here that a density barrier is created. Unless wind or currents break down the layered temperature and density structure, decaying organic matter that sinks beneath the region of change cannot pass back through. Hence, while light is abundant near the surface, nutrients tend to be plentiful and trapped at depth. This is particularly the case in the tropics and subtropics, where a thick layer of warm surface water almost always floats above the cold deep ocean. As a result, the surface waters of the tropics tend to be relatively unproductive and thus very clear. In the cold, windy polar regions, the temperature difference between the surface and deep water is minimal, so phytoplankton growth is usually limited only by light availability. In the midlatitudes, where seasonal variations are strongest, the sea's temperature layering is often a temporary phenomenon. Here the dividing line between warm and cold is more gradual and throughout the year changing conditions can result in periodic blooms of life.

Algae Blooms

The temperate zone is sunny, relatively calm, and warm in the summer and darker, windy, and cold in the winter. As a result, two distinct blooms of plant material commonly occur, one in the fall and one in the spring. In the summer months, the strong sun provides a wealth of energy for photosynthesis. Marine plants and algae grow rapidly, but they quickly use up all of the nutrients in the sea's surface waters, so productivity falls off as the summer progresses. With the approach of fall and winter, the sun remains relatively intense, and storms bring increasingly strong winds. Mixing by the wind breaks down the sea's temperature layering and density barrier, and brings nutrients up into shallower water. The phytoplankton receive a windfall of nutrients and there is an enormous explosion of life at the sea's surface, often referred to as the fall bloom. Not only do phytoplankton bloom, but with an abundance of food available, so do the zooplankton, and life in the sea prospers. During the winter, strong winds blow frequently and constantly mix ocean waters;

nutrients are plentiful, but now the sun's intensity fades, so productivity again declines. Come spring, light levels begin to increase. At first, nutrients in the sea are still abundant, so the photosynthesizers bloom once more. In general, the spring bloom tends to be somewhat smaller than that of the fall.

In the open ocean, algae blooms feed the larger creatures of the sea and typically pose little threat to others. However, similar blooms in the coastal ocean or inland bays can create conditions devastating to other marine life and potentially harmful to humans. Sewage, manufacturing wastes, and runoff from agricultural areas tend to contain high levels of nitrogen and phosphorus, which if dumped or carried into the sea where flow and mixing is restricted, may stimulate an algae bloom and create a situation commonly called *eutrophication.* When a bloom of algae is so rapid and intense that grazing animals cannot keep up with production, a dense mat of algae may form over the sea surface. The thick carpet of slime blocks off light and oxygen to the waters below and the underlying algae begin to die and decompose. Decomposition uses up the remaining oxygen and produces carbon dioxide, causing the bloom waters to become a foul and deadly soup. Without oxygen, fish and other creatures begin to die and the entire ecosystem deteriorates. In the deep Gulf of Mexico, in a region rich with commercial and recreational fisheries, an area the size of New Jersey has been christened the "dead zone." High levels of nutrients washing into the Gulf from the Mississippi and Atchafalaya river systems periodically create a zone near the seafloor devoid of or very low in oxygen. While fish and other mobile creatures can swim away from the deadly conditions, slow or sedentary organisms cannot. The dead zone appears to begin in late spring, reaching a maximum at midsummer, and fading in the fall. Following the 1993 Mississippi River flood, the dead zone doubled in size, reaching 18,000 square kilometers. Stable temperature layering within the water during the summer months, combined with excessive inputs of nutrient-rich water (possibly from the increasing use of fertilizers in the region), appears to be the cause of the dead zone's seasonality. Scientists are studying the dead zone and trying to determine how to prevent its occurrence in the future.

Massive algae blooms can also cause the water to turn various shades of brown or red, generating a phenomenon called *red tide*. The color of the water typically reflects the color of the principle species involved, usually dinoflagellates. In some cases, the species of algae that blooms can itself be toxic to other marine creatures, and can cause serious illness in humans. Of the roughly 5000 phytoplankton species, fewer than 80 are known to be toxic (NRC, 1999). For humans, harmful algal blooms threaten health when ingested in contaminated seafood or water, or inhaled in tainted sea spray. One extremely toxic type of algae, dubbed "the cell from hell," is the dinoflagellate *Pfesteria piscicida;* since its identification in 1991, it has caused massive fish kills, shellfish mortality, and raised public health concerns. During laboratory investigations of *Pfesteria,* researchers have reported skin lesions, lightheadedness, and loss of memory. There have been no human illnesses reported from eating fish exposed to *Pfesteria,* but scientists remain unsure of the health risks posed by seafood obtained in areas of related fish kills. Runoff containing wastes high in nitrogen and phosphorous is thought to promote the growth of *Pfesteria*.

Another suspect dinoflagellate can cause ciguatera, a type of poisoning that occurs when humans ingest contaminated fish. Ciguatera poisoning can lead to debilitating illness characterized by chronic pain and numbness in the arms, legs, joints, and bones, and oddly enough, a reversal of temperature sensation. Hot feels cold and cold feels hot. Symptoms can persist for weeks to months and, in a few cases, even years. Ciguatera strikes about 50,000 people each year, principally in the Caribbean, Florida, the Hawaiian Islands, French Polynesia, and Australia (NRC,

1999). The toxin associated with ciguatera appears to rise through the food chain and accumulate in large carnivorous fish. It begins with small grazing fish feeding on toxin-laced algae in a reef, and ends up in larger fish that consume the smaller fish. Cooking does not kill or alleviate the toxicity of ciguatera. Fish that have a propensity to carry ciguatera, such as large barracuda and some of the bigger jacks, should be avoided. As with other fish and shellfish toxins, scientists are working hard to develop a fast, economical, and efficient means to detect ciguatera in fish meat.

In many cases the mechanisms and processes that create harmful algal blooms are poorly understood. Worldwide, there has been a reported increase in the occurrences of harmful blooms, and some now suggest that they may be an indicator of regional and global environmental change (*Marine Ecosystems,* 1998).

Where the ocean contains little organic or inorganic matter it is colored deep blue. If marine plants are present, they tend to absorb blue light and reflect green, thus changing the ocean's color. Ocean color can also be influenced by the presence of suspended sediments and dissolved organic matter. Thus by taking a snapshot of the color of the ocean's surface water we can learn about plant abundance and productivity, runoff of sediment-laden river water, and even pollution. From 1978 to 1986, NASA operated the coastal zone color scanner (CZCS) and brought scientists the first large-scale images of ocean color. Because of the great success of the CZCS, NASA recently launched the next generation instruments to measure ocean color from space, the SeaWIFS satellite. The SeaWIFS (an acronym for sea-viewing wide field-of-view sensor) satellite is an exciting, relatively new tool to document and study algal blooms as well as the overall productivity of the sea. It measures violet, blue, yellow, and green hues in the ocean, as well as the intensity of red light scattered by dust and haze in the atmosphere and on land. Ocean color data is already helping us to understand the global distribution of marine plants; "hot spots," or blooms of biological activity; river input; and the ocean's role in the uptake of carbon dioxide (Color Plate 9). The measurement of ocean color from space is unquestionably providing a fascinating new means to learn more about the biological sea.

The biology of the ocean is a seemingly endless tale of nature's ingenuity and fantastic diversity. We have learned but a smidgen of the overall story, limited by our sampling ability and land-oriented way of life. One thing however, is crystal clear: there is much more to discover about the ocean's watery world and its vast array of creatures.

A Once-Bountiful Sea

More has changed in the ocean in the last half century than in all of human history, and the problem stems not only from what we are putting into the sea but what we are taking out.
—*Sylvia Earle*

*O*NCE UPON A time there was a sea in which tuna, stout with age, commonly roamed in schools not of 10 or 50, but hundreds. Huge aggregations of tuna chased even larger schools of small herring and mackerel, so thick with fish that from above they appeared as shallow ground. The lobsters in this ocean often reached some 2 meters in length and the cod grew as big as a grown man. Throughout this bountiful sea swam great pods of dolphins and whales. Seals and otters frolicked in abundance amid the waves, and an array of sharks fearlessly cruised the sea. Salmon, halibut, shrimp, and turtle were present in prolific numbers. To support such a profusion of life, the ocean's waters were rich with small floating plants and animals, while the bottom was alive with crawling, sedentary, and burrowing creatures. At the water's edge, the shoreline stretched for hundreds of kilometers, lined by little more than glistening sand, ragged rocks, and the shade of seaside trees. Tales recounting such a bountiful and wondrous sea are plentiful, but sadly, it is an ocean that we may never see again.

Florida Keys, More than Thirty Years Ago

Just south of Miami and the Everglades, a string of small islands arch to the south and west—the Florida Keys. The narrow pieces of land in the island chain are built of limestone, a remnant from a time past when the sea crested high above its current state. With very little commercial development, the islands are overgrown with mangroves and inhabited mainly by thick clouds of mosquitoes. Wading birds such as the pink spoonbill, egret, heron, and ibis come in large numbers to partake of the region's bounty, and the surrounding ocean is rich with marine life, a testimony to a still prosperous sea.

Along the shore, hundreds of creatures roam among the shadows flickering beneath the mangroves. Sponge, shellfish, and algae decorate the mangrove roots with fruitful growth, while young fish, crabs, and lobsters ramble under their tangled web. Spiny lobsters are everywhere—some so big that it is difficult for them to navigate the mangrove's maze of intertwining roots. When people start taking lobsters from the Keys, they can just walk the shore and simply pick them up. Many lobsters leave the mangroves and venture into deeper water to live within the crevasses of the nearby reefs and rocky seafloor; others crawl about the vast meadows of seagrass that lie close by.

In the seagrass beds, sponge, starfish, scallops, and small fish abound, while turtles and manatees often cruise by, feeding on the meadows of green grass that gently move in the waves. Fish and rays regularly tour the seagrass and benthic creatures busily burrow and sieve through the underlying sediments. Vast herds of conch slowly make their way across the bottom, jumping and hopping beneath the weight of their large, pink shells.

Where sand and rock lie bare on the seabed, the overlying water is filled with dense clouds of small, silvery fish. In a coordinated dance, they move gracefully as one within the sea. Larger fish linger nearby, waiting for the appropriate moment to zoom in for a tasty snack. Black, red, and nassau groupers are particularly abundant, lumbering wide-mouthed around the outskirts of the schooling minnows (Figure 51). Many of the groupers are large, 25 pounds or more. In every nook and cranny, black spiny sea urchins sit in tightly packed clusters. Their dark spines, long and narrow, are sometimes tipped with a speck of translucent white. On the reef, branching corals grow in dense tan thickets, while large head corals build massive overlapping brown and yellow mounds (Color Plate 13). Purple sea fans and whips slowly wave back and forth, and large, orange basket sponges sit quietly filtering the water. Colorful clumps and patches of algae speckle the bottom, though most of the reef's hard surfaces are kept clean by an abundance of grazing fish and sea urchins. Small fish constantly flutter in and out of the reef's shadowy framework decorated in a myriad of stripes, spots, and psychedelic patterns. A variety of sharks periodically swim by in search of prey. Florida's coral reefs are like a densely populated and affluent underwater city, teeming with colorful residents.

Figure 51 *The Nassau grouper.*
(Courtesy of National Marine Fisheries Service, NOAA.)

Some also tell of freshwater springs that bubble up in the shallow waters of Florida Bay, and of a frequent clarity in the water, rarely matched in later years.

Florida Keys, In the Year 2000

Today, the Florida Keys remain a spectacular marine setting and a wonderful place to explore the beauty of coral reefs or the shallow, wild backwaters of Florida Bay. Yet given stories of the past, the region is but a shadow of its past splendor. A large proportion of the mangroves have been removed or filled in to create land for construction. Most of the passageways connecting the Atlantic Ocean with Florida Bay have been infilled, and a flood of fast-food restaurants, hotels, and shops line the highway running the length of the Keys. Freshwater is piped in from mainland Miami, sewage wastes are pumped into the underlying, porous limestone, and hundreds of boats traverse the Keys' near-shore waters. Periodically, a plane flies low overhead spraying a cloud of chemicals to keep the mosquitoes in check. And several times each year, a large tanker or ship goes off course and crashes onto the nearby reefs.

Lobsters are no longer found under the mangrove roots but remain cautiously hidden in the dark crevasses of the seafloor, rarely getting larger than the legal catch size. Large crabs are seldom seen and few scallops and sponges inhabit the seagrass beds. One of the Keys' most enduring symbols, the conch, have all but disappeared. It is now illegal to collect conch in the Florida Keys. Seagrass beds have been reduced by disease, and scars from boat propellers crisscross the remaining meadows like open, bare wounds. Manatees are sighted infrequently and almost always bear deep scars down their wide backs from boat propellers. Turtles are becoming fewer and in Florida Bay, they often have massive clusters of bulbous tumors on their flippers and heads from a virus that is spreading through the population.

The once-abundant schools of minnows are now few and far between. The big groupers or other large, predatory fish rarely hover at their edges. It is an uncommon treat to come upon the curious grin of a nassau grouper of any size, or see a shark, for that matter. In many areas, thick mats or long strands of algae cover the rubble-strewn bottom. On the reefs, the bottom often appears enveloped by a plush shag carpet, having grown fuzzy with algae growth. The once-prolific black spiny sea urchins, *Diadema,* are few in number and seldom seen. In 1983 a deadly disease ravaged the *Diadema* populations in the Caribbean, Florida, and Bahamas. In the Florida Keys and throughout the world, corals are exhibiting frequent episodes of disease and bleaching, turning white as the symbiotic dinoflagellate zooxanthellae leaves their tissues. Large basket sponges and sea fans are also dying from disease. The near shore waters and those overlying the reefs are often more murky than in the past; few are the days of 30-meter visibility.

Some consider coral reefs the rainforests of the sea and indicators of global ocean health. Their degradation in Florida and around the world may indicate that our oceans are in trouble. But like the ocean as a whole, our understanding of how reefs change, grow, and die over time is relatively poor, so it is extremely difficult to distinguish between natural and human-made variability. Though our knowledge of reefs is incomplete, certain activities are undeniably detrimental to their well-being. Overfishing and destructive fishing practices (cyanide, dynamite, and bleach), unmanaged dredging, boat groundings and anchoring, coral collecting, the release of polluted runoff or discharges, and global warming are all damaging reefs throughout the world.

While Florida illustrates the changing nature of the sea, it also exemplifies a region where people are trying to effect positive change. Within the Florida Keys, residents and visitors alike are becoming increasingly aware of the link between the economy and health of the region with the well-being of the ocean. Thus, there is growing support for marine conservation and protection. Much of the commercial (gamefish and bonefish) and recreational fishing in the region is now strictly catch and release. Within the dive industry, operators use mooring buoys rather than anchoring, promote reef cleanups, and teach tourists environmentally friendly diving and snorkeling techniques. Educational programs and nonprofit organizations are intensifying their efforts in the Florida Keys to improve public awareness and appreciation for the ocean. The government and private foundations are spending more money to study Florida's reefs, determine the causes of deterioration, and protect the marine environment. In recent years a new understanding has developed about studying environments not as single entities but as links within a larger, integrated ecosystem, and this is no better exemplified than in South Florida. Today, government agencies, scientists, engineers, environmental leaders, ocean managers, and many others are working together in an unprecedented effort to restore not just the reefs, the Everglades, or Florida Bay, but all these regions as they are linked together as the South Florida ecosystem.

Another arena in which Florida exemplifies efforts to restore and preserve the ocean environment is marine protection. Florida Bay is part of Everglades National Park, and the nearshore waters and reef along the entire length of the Keys are now part of the Florida Keys National Marine Sanctuary. The establishment of a Keys-wide marine sanctuary initially met strong opposition, despite growing evidence of a deteriorating sea. It was a hard-fought battle to convince many that a marine sanctuary would be of benefit to all. Treasure hunters, jet skiers, sport and commercial fishermen, and others feared government control and constraints on their freedom to use the sea. Even today, placards and signs are still visible which read, "Say No to NOAA," the National

Oceanic and Atmospheric Administration, the agency that oversees the National Marine Sanctuary Program. While extensive wilderness areas on land are protected through our national park system, today there are just 12 marine sanctuaries in the U.S., which in total protect only about 1 percent of our nation's underwater wilderness. And although the entire length of the Florida Keys is now part of a marine sanctuary, actually less than 2 percent of its area is an actual "no take" reserve, where all fishing and collecting is prohibited. Visitors are not excluded from the marine reserve areas, but upon departing the undersea realm they may take with them only photographs and memories, which will last a lifetime. Scientific research has shown that in no-take areas, fish and other organisms multiply relatively rapidly and can begin to replenish diminished species (Bohnsack, 1994). However, despite such evidence, efforts to expand no-take areas in the Keys continue to meet powerful opposition. A recent proposal to make 20 percent of the Sanctuary no-take met with such contention that leading proponents were literally burned in effigy. Today, a valiant effort is being made to increase the no-take area in the Keys, to improve management practices, and to further restore and protect the surrounding ocean.

Florida is but one example of the many regions whose past tells of a bountiful sea now altered by human development and unmanaged use. Fisheries off the New England coast tell the same tale, so well exemplified by the disappearance of the cod and the closing of Georges Banks to fishing, once one of the most productive fishery grounds in the world. The Pacific Northwest's salmon have suffered a similar fate, along with the world's whales, seals, and otters. And sharks, so maligned by the public, are now being hunted down till their numbers dwindle. One of the most shocking practices is that of shark finning. The demand for high-priced shark-fin soup, considered a delicacy in Japan, has created a profitable market for shark fins, so fishermen catch sharks by the thousands, slice off their fins and throw them back into the sea, where they die a slow and terrible death. The awesome tuna, the swimming elite of the sea, is also swiftly declining in numbers because of overfishing. It is difficult to curtail fishing for tuna when just one fish can bring such a high price: a single bluefin tuna can net $25,000.

In the past, tuna were fished commercially with poles, pulling in one massive fish at a time. Today, the tuna industry relies on enormous nets set around floating objects under which the tuna tend to aggregate. Such practices capture not only immense numbers of adult tuna at one time; they also inadvertently and fatally ensnare an abundance of other creatures. For every 1000 tuna nets set out, some 654 billfish, 102 sea turtles, 13,958 sharks, and 2 dolphins are killed (Safina, 1998). In addition, many juvenile tuna are caught, affecting the population's ability to rejuvenate. Nontarget organisms captured while fishing are called by-catch and are usually unceremoniously dumped dead or dying back into the sea. In shrimp trawling, possibly the worst by-catch offender of all, for each pound of shrimp kept by a trawler, two to four times as much may be killed and disposed of overboard (Safina, 1997). Not only are we depleting the sea of species we use commercially, but we are also killing those that have no use to industry. Meanwhile, the number of "useful" species continues to rise; with the advent of the fish stick and other processed fish products, more types of fish are being eaten by more people (Kurlansky, 1997). Even the loglike sea cucumber is not safe from human depredations. In the Galapagos Islands, a once-thriving population of sea cucumbers has been drastically reduced, because they too have become a commodity in Japanese markets. The list of dwindling ocean creatures goes on. Within the United States alone, more than 80 percent of the 191 commercial fish stocks are fully exploited or overfished (NRC, 1998). Overfishing also affects the sea's own, for every fish or crustacean or sea urchin removed from the ocean affects the web of life. When we net krill or fish, we take food from the whales or sharks, when we collect abalone and sea urchins, we steal the otter's bread, and so on; our impact goes far beyond our own resource needs.

Fishing itself is not the problem, but fishing in numbers beyond what the ocean can sustain is. Although many would argue about the definition of sustainable fishing, it takes little more than logic to recognize the problem and some of its causes. The number of people inhabiting Earth continues to escalate and with the burgeoning population comes the increasing need for food. Around the world, some 20 percent of the animal protein consumed by humans comes from fishery products (FAO, 1995), but in contrast to the human population, fisheries are not growing, they are declining. There are simply too many mouths to feed and too few fish. And although aquaculture is on the rise, estimates are that it alone cannot provide the additional amount of fish needed to satisfy the growing demand for seafood (Safina, 1998).

With more people to feed, there are more people fishing the sea, and with improved technology, they are doing so more effectively than ever. Radar and GPS have greatly improved the navigation ability of fishing vessels, and planes as well as satellite imagery are now used to identify schools of fish or favorable ocean fronts where they may congregate. Fishing gear is being deployed in almost unthinkable proportion. Longliners may set over 100 kilometers of line equipped with thousands of baited hooks. Trawlers may drag a net large enough to engulf 12 jumbo jetliners, and though illegal in most places, some boats still use 64-kilometer-long drift nets (Safina, 1998). One of the most surprising aspects of overfishing is government support. On an annual basis, the fishing industry recently spent approximately $124 billion to catch just $70 billion worth of fish; the $54 billion deficit is made up by subsidies (Safina, 1998). In a goodhearted, but misguided effort to preserve jobs in the fishing industry, governments worldwide provide subsidies such as fuel-tax exemptions, price controls, low-interest loans, and grants for gear or boats. But with too many people using too many well-equipped boats, the fishing industry is literally self-destructing. We must abandon short-term crisis responses and consider long-term solutions to the fisheries dilemma. Quite simply, we must find ways to eliminate overfishing, reduce by-catch, and identify and protect areas in the oceans, where fish live and breed.

It may be that we can no longer commercially hunt the wild fish of the sea. Both Safina and Earle have put it eloquently in a simple comparison between land and sea. As Carl Safina ponders in the *Song for the Blue Ocean*, "a last buffalo hunt was occurring on the rolling blue prairies of the oceans." In *Sea Change*, Sylvia Earle writes, "Many who would recognize the absurdity of a plan to sustain large and growing numbers of people by hunting and gathering from the land buffalo, deer, wild birds, rabbits, squirrels, roots, and berries seem to disengage their power of reason when it comes to the sea." Some call commercial fishing a harvest, but what fishermen of the sea have planted and sowed their crop. We are simply taking, not replenishing, the ocean's wild stocks. On land, hunting is limited to recreational or subsistence living purposes. Yet in the ocean, we continue to hunt wild animals commercially on a massive and nonsustainable scale.

The influence of humankind on the ocean extends way beyond overfishing. Development along the coast is altering the flow and dynamic nature of shoreline environments and reducing the habitat or living space for marine communities. Oil and chemical discharges and polluted runoff are reducing water quality and altering nearshore waters. The fallout of mercury, nitrogen, dust, and other substances from the atmosphere, as well as global warming and ozone depletion, may also be causing the deterioration of marine environments. New studies show that incidences of harmful algal blooms and disease are on the rise and that the introduction of exotic species, particularly via ship ballast waters, is increasingly a threat to natural marine populations. Every hour an average of more than 2 million gallons of ballast water, meaning 2 million gallons of foreign plankton, are released into U.S. waters (Carlton et al., 1995). Zebra mussels have invaded the Great Lakes, Asian copepods and crabs have appeared along North America's eastern coasts, and a Black Sea jellyfish has been found in San Francisco Bay. Australian waters are being invaded by millions of European green crabs, an aquarium-bred algae is spreading unchecked through the Mediterranean, and in Russia's Sea of Azov, an American comb jelly is devastating the anchovy fisheries. We have dramatically altered the ocean's web of life and a once-bountiful sea has turned to troubled waters.

The problems in the ocean are becoming abundantly clear; the sea can no longer be considered limitless, inexhaustible, or impervious. A major challenge for the 21st century will be to find and implement effective solutions to the problems we have created in the ocean. We have already begun to implement federal and state programs to conserve fisheries, protect water quality and habitats, rebuild endangered species, and guide environmentally sensitive development. But as we learn more about the sea and our own impact, programs must be updated and additional efforts and investment made. The reality is that in Florida and around the world there will always be many competing needs and interests for ocean resources, and just as many opinions on how to manage their use and protection. Ocean and fisheries management is an extremely difficult, contentious, but critically important endeavor. To preserve the sea is, simply, to preserve life on Earth.

Issues in Marine Policy

The ocean community and the nation face many challenges in the upcoming century with regard to marine policy. If we are to continue to rely upon the sea's wealth while restoring and protecting its bounty, such obstacles must be overcome. As flow in the air and sea make for one global ocean, so too do they join one nation to another. By its very nature, the sea, covering nearly three-quarters of the planet, creates one global ocean community, linking all nations. Today, territorial waters can be delineated by highly precise satellite-based positioning, but currents, fish, marine mammals, and ships continually cross the sea's invisible borders. Thus, in addition to local, regional, state, and federal concerns, marine policies must be developed and implemented in an international context.

> *We have the power to damage the sea, but no sure way to heal the harm.*
>
> *—Sylvia Earle*

Our very way of thinking as terrestrial beings must be shifted to one that reflects a society surrounded and sustained by seawater. Simply, the oceans must become a national as well as a personal priority. Many would say that even on land we have not managed or protected our resources well. If this is true, then let us learn from our mistakes and prevent similar occurrences in the sea. On land, tradition and history have created a culture based on private ownership. With regard to the ocean, we must go against our time-worn notions of privatization and think in terms of a public domain, the collective good. Can one person or company own a plot of the open ocean and control the fish or flow within its boundaries? In some regions, we have tried to apply private ownership to beachfront property, but when storms wash away the sand or longshore currents erode a beach, the public (the taxpayers) is the one to pay.

Another problem in our way of thinking is how we apply value-based decision making. Without viable alternatives we often rely on short-term cost-benefit analysis to drive policy decisions. Yet how do we put a market value on the ocean? Easy to articulate, but difficult to quantify. How much is it worth to spend a day on the beach or go fishing? What is the value of the ocean's role in climate or the diversity of life on Earth? In 1997, the value of the ocean's resources was estimated at $21 trillion (as compared to $12 trillion for land-based resources); this valuation included food, recreation, protection from storms, shipping, waste removal, water recycling, and wildlife habitat (Costanza et al., 1997). But there is great uncertainty in such numbers, particularly considering that the ocean remains relatively unexplored and poorly understood. We are attempting to manage and make decisions about a system that we are only just beginning to know. Managing fisheries is a difficult task if you do not know how many fish there are, where they live and breed, and how their populations naturally change with time. And since individual marine organisms live within a system whose parts are so closely intertwined, it is risky to manage one species without understanding its impact and relationship to others and the surrounding physical, chemical, and geologic environment. But we cannot wait to have scientific certainty before acting, so policymakers and managers must push forward despite uncertainty and incomplete knowledge. And time is running out.

Because the ocean's wealth of resources is so diverse, our interests in the sea are similarly diverse and often competing. We look to the ocean for food, commerce, transportation, security, recreation, health, and environmental well-being. In the United States alone, one out of every six jobs is marine-related, fisheries add about $20 billion to the economy, and coastal destinations are the fastest-growing sector in the tourism and recreation industries. We rely on ports for most of our foreign trade, the seabed for oil, gas, and minerals, and the ocean waterways to secure national freedom, security, world peace, and humanitarian aid. The seas are the playing field upon which we fight a battle against drug smugglers and illegal immigration. And new discoveries in the ocean are beginning to reveal the sea's potential in the pharmaceutical and biotechnology industries. Such a broad community of ocean users and interests leads to competing needs and even difficulty in just communicating between people with varying perspectives.

Just as the vast, inhospitable, and opaque nature of the sea has kept its scientific secrets hidden from human eyes over the ages, these same properties offer obstacles to effective management of its resources. For instance, the enforcement of ocean regulations is extremely difficult given the sea's enormity and often harsh conditions. Can we find more effective means to identify and monitor those who pollute in the remote open sea or along our lengthy shores?

Other important issues in marine policy are a lack of national investment, national strategy, and coordination among regulating and funding agencies. The ocean sciences today represent less than 4 percent of the total federal research budget, down from 7 percent only 15 years ago. The amount of national investment in ocean science, exploration, protection, and technology development is embarrassingly small, particularly when compared to that of space exploration. This is not meant to demean the importance of the space program, but to encourage the public and our representatives in government to give equal weight to exploring and understanding the oceans, which make up most of our own planet's surface and sustain life as we know it. Although there are numerous agencies in the government that have ocean-related responsibilities and interests, there often is little coordination and cooperation between them. Just look at the list of some of the agencies involved:

- National Oceanic and Atmospheric Administration (NOAA), which includes the National Weather Service, National Ocean Service, National Marine Sanctuary Program, National Marine Fisheries Service, and the National Undersea Research Program
- National Aeronautics and Space Administration (NASA)
- Department of Interior, which includes the Geological Survey, Fish and Wildlife Service, National Park Service, and Minerals Management Service
- Environmental Protection Agency (EPA)
- National Science Foundation (NSF)
- Office of Naval Research (ONR)
- Naval Oceanographic Office
- U.S. Coast Guard (USCG)
- Department of Transportation
- Department of Energy
- Department of State

Once again it is a good news, bad news story. With such widespread involvement in ocean-related issues, a broad range of government resources and expertise are available. However, with little coordination there can be duplication, miscommunication, difficulty in accomplishing multidisciplinary and multi-agency projects, and overregulation, let alone a lack of a collective voice for the oceans. Undoubtedly to take advantage of the benefits of the current system we need better coordination among agencies, and where appropriate, a reduction or elimination of overlapping responsibilities.

In 1998, as part of the United Nation's International Year of the Ocean, the Department of Commerce and the Department of the Navy cohosted the National Ocean Conference in Monterey, California. The intent of the conference was to highlight the role of the oceans in everyday life, raise public awareness of ocean issues, and foster an effective dialogue among the members of the wide-ranging ocean community. It was the first time leaders from government, industry, academia, environmental groups, and other nongovernmental organizations all came together to discuss national ocean issues and solutions. Fishermen spoke with environmentalists who spoke with regulatory agency representatives who spoke with scientists. Discussion was clearly fueled by a passion for the ocean and highly varying perspectives on and interests in the sea. A report from the conference entitled *National Ocean Conference: Oceans of Commerce, Oceans of Life* gives an excellent overview of many of today's most significant marine issues and is available from the U.S. Department of Commerce, National Oceanic and Atmospheric Administration, Office of Public and Constituent Affairs.

One resounding message from participants at the National Ocean Conference was that the United States should, without hesitation, join the 1982 U.N. Convention on the Law of the Sea and become an official member of the international legislative arena for the oceans. The Convention on the Law of the Sea and the accompanying 1994 Seabed Mining Agreement address issues such as military and commercial navigation, fishing, oil and gas development, offshore mining, and scientific research. Every person at the ocean conference addressing the issue agreed that not joining puts the United States at a distinct disadvantage in the international arena of ocean management and undercuts our nation's role as a leader in the global ocean community. Some 50 nations, including most of our NATO allies and all other permanent members of the U.N. Security Council, are party to the convention—but not the United States. It appears that one member of the Senate is responsible for our failure to join the world in an endeavor to promote wise use and protection of the oceans. This one man has single-handedly blocked any Senate action on the U.N. convention. The public and our political representatives must step forward and take a proactive stance on the Law of the Sea issue if it is to be resolved soon.

Several other themes cut across many of the discussions, reports, and sideline conversations at the National Ocean Conference; these included the need for improved science education and training in a technologically sophisticated world, more partnering between entities such as government, industry, and academia; and a greater national interest and investment in the ocean. Conference participants all agreed that our nation needs a national strategy to define where we are headed and how we are going to get there when it comes to the ocean. Currently, the Oceans Act has 39 cosponsors in the House of Representatives and 19 cosponsors in the Senate. If approved, the Oceans Act would establish a commission to recommend an integrated and effective national ocean policy and advise the president on ocean issues. Whatever the fate of this legislation, it is clear that some means is needed to improve our national strategy and policy toward the sea.

One of the most successful aspects of the National Ocean Conference was simply bringing the wide community of ocean constituents together and facilitating an open, enlightening, and productive discussion. But while discussion among leaders of the ocean community and our nation is important, it is just as critical that the broader public becomes more ocean literate, more ocean-caring, and more outspoken on ocean issues.

WAVES OF THE FUTURE

We've made the investment needed to venture into the skies, and it has paid off mightily. We've neglected the oceans, and it has cost us dearly. This is the time to do for the oceans in the 21st century what our predecessors did for space.

—Sylvia Earle

Frontiers of Ocean Science

TODAY, THE OCEAN remains a vast frontier whose mysterious darkness, abundant life, and influence on Earth are just beginning to be known. In fact, much of what we have learned about the oceans is just how little we really know. Our ability to sample the sea and study its great complexity has improved dramatically in the last century, and the future is rich with the promise of exciting discoveries and practical purpose.

Technologically we have come so far, so fast, that it is difficult to even imagine what the future will bring. Satellite instrumentation is providing never-before-seen large-scale images that tell of sea surface height, temperature, ocean color, surface waves, and winds. Satellite tracking is allowing us to follow flow within the sea and the movement of its creatures as never before. And space technology now provides new data about the atmosphere and climate and allows for unprecedented precision in geographic positioning. Undoubtedly as our interpretation of data from satellites improves and as new instruments are developed, spaceborne study of the sea will intensify and expand.

Today, we also have a new view of the ocean, one that goes beyond the capabilities of the human eye. The study of microbes, the bacteria and viruses that inhabit the sea and its underlying crust, is one of the hottest and most rapidly expanding fields in ocean science. DNA and molecular technology have and will continue to enhance our ability to sample, identify, and study the sea's smallest creatures—its microbial population. In every teaspoon of seawater there may be more than 1 million microscopic organisms (DeLong and Ward, 1992). New reports are revealing the links between the ocean, marine microbes, and waterborne disease, as well as the great potential for new medicines and biotechnology derived from marine bacteria and compounds. Scientists have recently discovered that the sea's zooplankton can play host to the bacteria responsible for cholera and that outbreaks of the disease in Bangladesh are linked to changes in sea surface temperature. We may be able to predict potential cholera outbreaks from images of ocean color, which reveal plankton blooms. With our improved ability to study the microbes of the sea, we may find that they play a more dominant role in the marine environment, human health, and Earth than ever imagined.

Much of our present-day understanding of the ocean is built on a past in which the military invested heavily in marine science and technology. Today, with the cold war at an end, the ocean is less a medium of warfare than a place of environmental concern. For ocean science the end of the cold war is both good and bad news. The bad news is that with less national investment in ocean science it is more difficult to develop undersea technology and study the sea. The good news is that a wealth of ocean information and equipment is now becoming available to civilian scientists. The declassification of military data is allowing researchers access to information long secreted in government files. Ships and other instruments that were once used solely for military purposes are now becoming available for scientific research. For instance, a global array of underwater microphones once used to detect and monitor submarine activity is now being used by scientists to study hydrothermal vent activity, whales, and changes in global ocean temperatures. Using the Navy's sound surveillance system (SOSUS), civilian scientists can listen for and locate volcanic eruptions on the seafloor or detect migrating whales. In addition, because the speed at which sound travels through the ocean depends on temperature, scientists have been using the SOSUS array to study changes in the temperature of the global ocean. The use of sound to study the ocean and seafloor arose early in our endeavor to explore and understand the sea. Most assuredly, acoustic technology will continue to provide an exciting avenue for ocean research and exploration.

Another exciting, relatively new arena for ocean science is real-time monitoring combined with easy and widespread access to data. Today, almost everyone has access to timely weather updates and forcasts via television, newspaper, radio, or the Internet. Weather forecasting and reporting is partly possible because we have access to phenomena as they happen; cloud formation, rain, wind, and air temperature are all monitored in real time. In the ocean, we are just beginning to build systems capable of similar monitoring and communicating in real time. In fact, many believe that putting together an integrated and coordinated global ocean observing system is the highest priority in ocean science today. Several smaller scale systems are already in operation and illustrate the use and benefits of real-time ocean monitoring.

In 1997, for the first time, the ocean observing system in the Pacific allowed us to recognize and forecast an impending El Niño event 6 months before it was in full swing. With the prediction of future El Niños, farmers can prepare crops, engineers can shore up areas vulnerable to erosion, and fishermen and ocean farmers can plan for changing fish populations and seawater temperatures. In the coastal zone, in places such as New York Bight, Chesapeake Bay, the Texas shelf, and Monterey

Bay, real-time monitoring is also providing timely and widespread access to changing ocean conditions. And to assist navigation and safety, NOAA operates PORTS, a physical oceanographic real-time system in Tampa Bay, New York, and San Francisco Bay. Through phone lines, radio, and the Internet, scientists, ship captains, recreational boaters, fishermen, local officials, and other interested parties can access the PORTS system and learn about actual water levels, currents, winds, and other oceanographic and meteorological phenomenon.

Scientists are also creating remote undersea research stations to provide real-time access to the deeper sea. One such undersea station is LEO-15, a long-term environmental observatory located in 15 meters of water, 9 kilometers off the coast of New Jersey. LEO-15 consists of two remote seafloor observatories that are linked by a cable to the Rutgers Marine Field Station in Tuckerton, New Jersey. Through the Internet, researchers, educators, and the public can access the undersea world in near real-time and learn about a wide array of ocean processes. Instruments at the site measure water temperature, salinity, oxygen concentration, light, and chlorophyll levels. Continuous video of the bottom and measurements of sound, water flow, and waves are or will be available. Additionally, a small, autonomous underwater vehicle is being garaged beneath the sea at LEO-15 and will be used to examine episodic events that occur away from the actual undersea station. On command from a control station, the vehicle is launched, negotiates a previously programmed track, and then returns to dump its data.

The Hawaii Undersea Geo-Observatory (HUGO) is an automated underwater observatory recently installed on top of the Loihi seamount. HUGO is located some 80 kilometers south of Hawaii and connected via an AT&T junction box and cable to a shore station on the big island of Hawaii. Scientists will use instruments in the submarine station, including seismometers, an underwater microphone, and a bottom pressure sensor, to study and monitor such things as earthquake activity and seafloor volcanic eruptions. The station is also directly linked to the U.S. Geological Survey's Hawaiian Volcano Observatory and the Pacific Tsunami Warning Center, and data is available over the Internet. Another remote long-term seafloor observatory is about to be established; the New Millennium Observatory (NeMO) is being built on the crater of a huge undersea active volcano along the Juan de Fuca ridge some 480 kilometers off the coasts of Oregon and Washington. One purpose of NeMO is to better understand the community of microbes that lives beneath the seafloor and thrives at extremely high temperatures at active ridge sites. LEO-15, HUGO, NeMO, PORTS, and the Pacific monitoring system are a whole new and exciting way of doing business in ocean science. These endeavors not only provide traditional researchers and ocean users with unparalleled access to changing ocean conditions as they occur and over the long term, but also allow for the development of new and innovative ways of teaching about the sea—a virtual underwater classroom.

For years, the ocean's great expanse, depth, and darkness hid the largest mountain chain on Earth, an entirely new deep-sea ecosystem, and many of its creatures, like the coelacanth and megamouth shark. Our investment in marine technology, research, and exploration allowed us to uncover these and other secrets of the sea, and unquestionably many more remain to be discovered. Continued advances in remotely and autonomously operated underwater vehicles will certainly further exploration, but many would argue that there is nothing more valuable than first-hand human experience. Today, there are just five manned submersibles operated by four countries that are even capable of going to half of the ocean's depth and only one operating undersea laboratory. In the past, there were numerous submersibles and a multitude of underwater habitats. Recently, the Navy decommissioned *Sea Cliff,* the only submersible supported solely by the U.S. government.

We spend billions of tax dollars putting humans into space and potentially onto other planets, but ignore similar exploration of the ocean. Is it that the public falsely believes that we have nothing new to learn when it comes to the sea? The outlook for undersea exploration and ocean science is exciting, but tentative, having a surprisingly low national priority and too little investment.

What other topics will be on the forefront of ocean science? A 1998 report by the National Research Council's Ocean Studies Board highlights three areas that most agree will be the focus of research efforts far into the future: improving the health and productivity of the coastal oceans, sustaining ocean ecosystems for future generations, and predicting ocean-related climate variations over a human lifetime (NRC, 1998). We now recognize the significance, complexity, and changing nature of the connections between the ocean, atmosphere, and Earth. Today, not only do scientists study long-term change, but there is a growing emphasis on shorter-scale variability and determining how much of this variability is human-induced. Some believe that the most important achievements in ocean science in the future may come not from new advances in technology but from our ability to use what we already have in an integrated, collaborative, and productive way. The future for ocean science depends heavily on our ability to build partnerships among federal agencies, academia, industry, nonprofit groups, and other members of the international ocean science community. In addition, we must teach and inspire young and old alike to learn and care about the sea. As a nation and as inhabitants of this blue orb, we must make the ocean a true priority, providing adequate attention to and investment in its future, and our own.

Looking Back to Look Ahead

Sylvia Earle

NEAR THE END of the 19th century, history was made as scientists aboard the research vessel HMS *Challenger* set sail on the first-ever global voyage of ocean research and exploration. Equipped with a great assortment of bottles, nets, trawls, hooks, and other instruments, they traveled nearly 112,000 kilometers over three and a half years, documenting and later reporting on the nature of the sea as it was then known. What they discovered stunned the scientific world—thousands of new species; significant new insights about currents, sediments, and temperature; and new knowledge of exotic cultures and distant lands.

Now, as the 20th century comes to a close, it is tempting to speculate about what might most surprise the *Challenger* scientists if they could magically be transported forward to the present time. What would they think of airplanes flying overhead, of automobiles and atomic power, flights to the moon, and satellites used to survey the sea from space? Of plastics, microchips, and megaships, of computers, videos, GPS, and supermarkets filled with food and goods gathered from around the world? But mostly, what would they think of the revolutionary discoveries that have been made in the sea?

Based on their soundings in the 1800s, those aboard the *Challenger* may have suspected the existence of many more undersea mountains than had previously been discovered, but could they possibly have imagined that 60,000 kilometers of ridges, ranges, and peaks would be charted in the decades ahead? Or that the continents are in constant motion, their actions linked to processes originating deep beneath the sea? That life occurs throughout the entire ocean, from its sunlit surface waters to its greatest, deepest depths, and that the ocean contains vast communities of microbial creatures that provide important clues to the very origin of life itself?

What would have run through their minds if they could have read this volume about the oceans or discussed it with the author? Surely their reaction would be laced with excitement as well as astonishment considering that the author is a distinguished field scientist who has personally spent years studying the sea, working off boats, diving, living in undersea laboratories, and exploring in small submersibles. For a young woman to be engaged in such endeavors or become an accomplished scientist was unthinkable at the time of the *Challenger.* But as one astute observer has noted, "Half the fish, dolphins, and whales are female. Why not half the oceanographers as well?" Scientists from ages past would surely wish that they too might see and do even a fraction of what is now possible for anyone who wants to study the sea as a new era of ocean exploration and study unfolds.

No doubt they would be tremendously interested in the 20th-century technologies used for ocean exploration: acoustic systems for finding fish or mapping the undersea terrain, nuclear submarines, offshore oil platforms, undersea stations, the many-ton bathyscaphe *Trieste* that took two men to the ocean's greatest depths in 1960 for the first and only time and then returned them safely to the surface. Not to mention the hundreds of sophisticated, remotely operated vehicles that now ply the ocean, bristling with cameras and sensors. . . .

Yet of all the amazing revelations, two might stand out above the rest. First, despite the enormous strides we have made in the past century toward understanding the nature of the sea, fully 95 percent of the ocean remains to be explored. Given the magnitude of what our modest ventures have revealed to date, and given how much more remains to be found, *Challenger* scientists might well stand in awe with us wondering what else is down there. Can we even make an intelligent guess at what it might be? And given the tremendous resources and technological know-how that has been devoted to aviation and aerospace, why has so little attention been paid to the depths below? What would we know if the commitment to ocean research and exploration were on par with space exploration?

The second stunning revelation is the impact of humans on the ocean in the brief span of a century. Anyone who knew the ocean as it was 100 years ago would have to wonder what terrible events had occurred to so drastically alter its very nature. Even half a century ago, the ocean was a "sea of Eden." Anthropologist-explorer Thor Heyerdahl remarked, when he returned in 1947 from his famous *Kon Tiki* expedition across the Pacific that what had impressed him the most was not the new discoveries made, not the wondrous encounters with sea creatures, not even the excitement of staying afloat for many weeks in his improbably fragile balsa-log craft. Rather, his most memorable and unexpected discovery was how pristine the ocean was, how it must have been then, much as it was thousands, even millions, of years before. Great sharks and other fish swam up to the raft like curious children. None had ever encountered a terrestrial primate before. Nor had they seen the other strange newcomers that soon would become a constant reminder of changing times in the sea: drifting plastic cups, bags and bits of styrofoam, weathered tar balls, and fragments of discarded nets. By the time of Heyerdahl's 1970 expedition aboard the reed raft *Ra II,* there were clear signs of trouble. Not a day passed, he said, without sailing through windows of debris, sometimes great rafts of it, even hundreds of miles from shore.

Since the 1970s, the very nature of the ocean has changed as a result of what we have put into it. We have deliberately dumped all manners of waste into the sea and turned our backs while runoff laden with nitrates, phosphates, pesticides, herbicides, and other substances have flowed from fields, farms, backyards, and golf courses. Materials have been lofted high into the skies above that eventually return, falling onto the land and into the sea. Concurrently, hundreds of millions of tons of wildlife have been taken out of the ocean. As a consequence, populations of dozens of creatures once thought to be infinitely resilient have collapsed, from North Atlantic cod to Pacific rockfish and Indian Ocean tiger prawns. In the process, ancient systems, fine-tuned over hundreds of millions of years, have suffered great, perhaps irreparable, harm.

Now that we know the significance of the living ocean to the survival and well-being of humankind, there is a real and growing concern about the consequences of the recent, dramatic changes in the very nature of the sea. Increasingly, ocean research and exploration are being viewed as vital endeavors, no more a luxury born of curiosity, but essential if we are to understand what might be done to reverse the alarming trends in the sea. For the land, measures were taken at the turn of the last century to establish protective policies—and protected places—to safeguard wild systems that were threatened by pressures from human actions. From a modest beginning—the establishment of Yellowstone Park in the same year that *Challenger* set sail—grew an ethic of caring reflected in the National Park System. Some called it the best idea America ever had. As a new century approaches, a young but promising counterpart for the oceans, a growing network of marine sanctuaries, is underway. In the United States there are 12; worldwide, there are about 1200. Much remains to be done to insure that the ocean's wilderness and its creatures will continue to thrive into the next century and far beyond. Establishing and maintaining protected places in the sea is surely a move in the right direction.

As we reflect on what oceanographers from the HMS *Challenger* might have thought about our knowledge of the ocean and the state of the sea today, so might we consider our actions now and in the future, and how they might be viewed from afar. Efforts to study the sea have unlocked an unimaginable wealth of knowledge, and at the same time revealed both our ignorance and potentially harmful influence. Today, the world we live in is a reflection of decisions made in times past. What we do—or do not do—at this point of history will determine the fate of this ocean planet, and of humankind.

Sources and Suggested Readings

Baker, P., and M. McNutt. 1998. *The Future of Marine Geology and Geophysics: Report of a Workshop, Ashland Hills, Oregon, December 5–7, 1996,* National Science Foundation.

Balling, R. C. 1993. "The Global Temperature Data: Research and Exploration," *National Geographic,* 9(2): 201–207.

Barghoorn, E. S. 1971. "The Oldest Fossils," *Scientific American,* 244(5): 30–42.

Bartlett, K., and D. Wexler. 1998. "News and Notes," *Geotimes,* 43(12): 6–14.

Boersma, P. D. 1999. "Tuxedo Junction," *Wildlife Conservation,* March/April: 27–31.

Bohnsack, J. A. 1994. "How Marine Fishery Reserves Can Improve Reef Fisheries," *Proceedings of the Gulf and Caribbean Fisheries Institute,* 43: 217–241.

Briggs, D. E. G., Erwin, D. H., and F. J. Collier. 1994. *The Fossils of the Burgess Shale.* Smithsonian Institution Press, Washington, D.C.

Broad, W. J. 1997. *The Universe Below: Discovering the Secrets of the Deep Sea.* Touchstone/Simon & Schuster, New York.

Brower, K. 1981. "A Galaxy of Life Fills the Night," *National Geographic,* 160(6): 835–847.

Bushnell, P. G., and K. N. Holland. 1989. "Tunas: Athletes in a Can," *Sea Frontiers,* January-February: 42–48.

Carey, F. G. 1973. "Fishes with Warm Bodies," *Scientific American,* 228(12): 36–44.

Carlton, J. T., Reid, D. M., and H. van Leeuwen. 1995. "The Role of Shipping in the Introduction of Non-Indigenous Aquatic Organisms to the Coastal Waters of the United States (other than the Great Lakes) and the Analysis of Control Options," *The National Biological Invasions Shipping Study.* U.S. Coast Guard and the National Sea Grant Program.

Carson, R. 1955. *The Edge of the Sea.* Houghton Mifflin, Boston.

———. 1951. *The Sea Around Us.* Oxford University Press, New York.

———. 1941. *Under the Sea Wind.* Oxford University Press, New York.

Carwardine, M., Hoyt, E., Fordyce, R. E., and P. Gill. 1998. *Whales, Dolphins & Porpoises: The Nature Company Guides.* Time-Life Books/Weldon Owen, Australia.

Chisholm, S. W. 1992. "What Limits Phytoplankton Growth?" *Oceanus,* 35(3): 36–46.

Cloud, P. E., Jr. 1968. "Atmospheric and Hydrospheric Evolution on the Primitive Earth," *Science,* 160: 729–736.

———, and A. Gibor. 1970. "The Oxygen Cycle," *Scientific American,* 273(3): 110–123.

Corso, W., and P. S. Joyce. 1995. *Oceanography.* Applied Science Review. Springhouse Corporation, Springhouse, Penn.

Costanza et al. 1997. "The Value of the World's Ecosystem Services and Natural Capital," *Nature* 387: 253–260.

Decker, R., and B. Decker. 1998. *Volcanoes,* 3 ed., W. H. Freeman, New York.

DeLong, E. F., and B. B. Ward. 1992. "Biological Oceanography from a Molecular Perspective," *Oceanus,* 35(3): 47–54.

Deser, C. 1996. "A Century of North Atlantic Data Indicates Interdecadal Change," *Oceanus,* 39(2): 11–13.

Dickinson, W. R., 1998. "A Revolution in Our Time," *Geotimes,* 43(11): 21–25.

Dybas, C. L. 1999. "Undertakers of the Deep," *Natural History,* 108(9): 40–47.

Earle, S. 1995. *Sea Change: A Message of the Oceans.* G. P. Putnam, New York.

——— and A. Giddings. 1980. *Exploring the Deep Frontier: The Adventures of Man in the Sea.* National Geographic Society, Washington, D.C.

——— and W. Henry. 1999. *Wild Ocean: America's Parks under the Sea.* National Geographic, Washington, D.C.

Ellis, R. 1996. *Deep Atlantic: Life, Death, and Exploration of the Abyss.* Lyons Press, New York.

———. 1998. *The Search for the Giant Squid.* Lyons Press, New York.

Erickson, J. 1995. *A History of Life on Earth: Understanding Our Planet's Past.* Facts on File, New York.

Food and Agriculture Organization (FAO). 1995. *The State of the World Fisheries and Aquaculture.* United Nations, Rome.

Gardner, J. V., Dartnell, P., Gibbons, H., and D. MacMillan. 1999. *Exposing the Seafloor: High-Resolution Multibeam Mapping along the U.S. Pacific Coast. USGS Information Sheet.* U.S. Geological Survey/U.S. Department of the Interior,

Gould, S. J. 1989. *Wonderful Life: The Burgess Shale and the Nature of History.* W. W. Norton, New York.

Guberlet, M. 1964. *Explorers of the Sea: Famous Oceanographic Expeditions.* Ronald Press, New York.

Hansen, J., Lacis, A., Ruedy, R., Sato, M., and H. Wilson. 1993. "How Sensitive Is the World's Climate?" *Research and Exploration,* National Geographic Society, 9(2): 234–247.

Haq, B. U., and A. Boersma (eds.) 1978. *Introduction to Marine Micropaleontology.* Elsevier Science Publishing, New York.

Harbison, G. R. 1992. "The Gelatinous Inhabitants of the Ocean Interior," *Oceanus,* 35(3): 18–23.

Hartmann, W. K., and R. Miller. 1991. *The History of Earth: An Illustrated Chronicle of an Evolving Planet.* Workman Publishing, New York.

Heezen, B. C., and M. Ewing. 1952. "Turbidity Currents and Submarine Slumps, and the 1929 Grand Banks Earthquake," *American Journal of Science,* 250: 849–873.

Hogg, N. 1992. "The Gulf Stream and Its Recirculations," *Oceanus,* 35(2): 18–24.

Humphris, S. E., and T. McCollom. 1998. "The Cauldron beneath the Seafloor," *Oceanus,* 41(2): 18–23.

Hutchinson, G. E. 1970. "The Biosphere," *Scientific American,* 223(3): 45–53.

Independent World Commission on the Oceans (IWCO). 1998. *The Ocean Our Future.* Cambridge University Press, Cambridge.

Karl, T. R. 1993. "Missing Pieces of the Puzzle," *Research and Exploration,* National Geographic Society, 9(2): 234–247.

Kennett, J. P. 1982. *Marine Geology.* Prentice-Hall, Englewood Cliffs, N.J.

Kious, W. J., and R. I. Tilling. *This Dynamic Earth: The Story of Plate Tectonics.* US Geological Survey/U.S. Department of the Interior, Washington, D.C.

Komar, P. D. 1976. "Tides," in *Beach Processes and Sedimentation.* Prentice-Hall, Englewood Cliffs, N.J., pp. 122–146.

Kurlansky, M. 1997. *Cod: A Biography of the Fish that Changed the World.* Penguin Books, New York.

Lamb, S., and D. Sington. 1998. *Earth Story: The Shaping of Our World.* Princeton University Press, Princeton, N.J.

Larson, C. E. 1998. "The Chesapeake Bay: Geologic Product of Rising Sea Level," *USGS Fact Sheet 102-98.* U.S. Geological Survey, The Department of Interior.

Levin, H. L. 1991. *The Earth through Time.* Saunders College Publishing, Orlando, Fla.

Lutz, R. A., and R. M. Haymon. 1994. "Rebirth of a Deep-Sea Vent," *National Geographic,* 186(5): 115–126.

Marine Ecosystems: Emerging Diseases as Indicators of Change. 1998. Year of the Ocean Special Report. National Oceanic and Atmospheric Administration and NASA, Global Change Program. Harvard Medical School. Boston, Massachusetts.

Martin, G. 1999. "The Great White's Way," *Discover,* 20(6): 54–61.

Mazzuca, L., Atkinson, S., Keating, B., and E. Nitta. 1999. "Cetacean Mass Strandings in the Hawaiin Archipelago, 1957–1998," *Aquatic Mammals,* 25(2): 105–114.

McCartney, M. 1996. "North Atlantic Oscillation," *Oceanus,* 39(2): 13.

McDonald, I., and C. Fisher. 1996. "Life without Light," *National Geographic,* 190(4): 88–97.

Monastersky, R., 1998. "The Globe inside Our Planet," *Science News,* 154: 58–60.

National Ocean Conference: Oceans of Commerce, Oceans of Life. 1999. U.S. Department of Commerce/U.S. Department of Navy, Silver Springs, Maryland.

National Research Council (NRC). 1999. *From Monsoons to Microbes: Understanding the Ocean's Role in Human Health.* Ocean Studies Board/National Academy Press, Washington, D.C.

———. 1998. *Opportunities in Ocean Sciences: Challenges on the Horizon.* Ocean Studies Board/National Academy Press, Washington, D.C.

———. 1998. *Sustaining Marine Fisheries.* National Academy Press, Washington, D.C.

Neelin, J. D., and M. Latif. 1998. "El Niño Dynamics," *Physics Today,* 51(12): 32–36.

Nybakken, J. W. 1993. *Marine Biology: An Ecological Approach,* 3d ed. HarperCollins, New York.

Nybakken, J. W., and S. T. Webster. 1998. "Life in the Ocean," *Scientific American Presents,* 9(3): 74–87.

Oceanus, 34(4) 1991/92. *Mid-Ocean Ridges,* Woods Hole Oceanographic Institution, Massachusetts.

Partridge, B. L. 1982. "The Structure and Function of Fish Schools," *Scientific American* 246(6): 114–123.

Pinet, P. R. 1992. *Oceanography: An Introduction to the Planet Oceanus.* West Publishing, New York.

Pirie, R. G. (ed.) 1996. *Oceanography: Contemporary Readings in Ocean Sciences,* 3d ed., Oxford University Press, New York, Oxford.

Pond, S., and G. L. Pickard. 1983. *Introduction to Dynamical Oceanography,* 2d ed. Pergamon Press, New York.

Prager, E. J. 1999. *Furious Earth: The Science and Nature of Earthquakes, Volcanoes, and Tsunamis.* McGraw-Hill, New York.

Press, F., and R. Siever. 1982. *Earth,* 3d ed. W. H. Freeman, San Francisco.

Raup, D. M., and J. Sepkoski, Jr. 1986. "Periodic Extinction of Families and Genera," *Science,* 231: 833–836.

Restless Earth. 1997. National Geographic Society, Washington, D.C.

Richardson, P. L. 1993. "Tracking Ocean Eddies," *American Scientist,* 81: 261–271.

Safina, C. 1997. *Song for the Blue Ocean: Encounters along the World's Coasts and beneath the Seas.* Henry Holt, New York.

———. 1998. "The World's Imperiled Fish," *Scientific American Presents,* 9(3): 58–63.

Schneider, S. 1989. "The Greenhouse Effect: Science and Policy," *Science,* 243: 771–781.

Schoff, T. J. M. 1980. *Paleoceanography.* Harvard University Press, Cambridge, Mass.

Seibold, E., and W. H. Berger. 1996. *The Sea Floor: An Introduction to Marine Geology,* 3d ed. Springer-Verlag, Berlin, Heidelberg, Germany.

Smith, W. H., and Sandwell, D. T. 1997. "Global Sea Floor Topography from Satellite Altimetry and Ship Depth Soundings," *Science,* 277(5334) 1956–1962.

Snelgrove, P. V. R., and J. F. Grassle. 1995. "The Deep Sea: Desert and Rainforest," *Oceanus,* 38(2): 25–29.

Stowe, K. 1996. *Exploring Ocean Science,* 2d ed. Wiley, New York.

Tricas, T. C., Deacon, K., Last, P., McCosker, J. E., Walker, T. I., and L. Taylor. 1997. *Sharks and Rays: The Nature Company Guides,* Time-Life Books/Weldon Owen, Australia.

Tunnicliffe, V. 1992. "Hydrothermal-Vent Commentaries of the Deep Sea," *American Scientist,* 80: 336–349.

U.S. Global Change Research Program. 1997. *Our Changing Planet: The FY 1997 U.S. Global Change Research Program.* Office of Science and Technology Policy, Executive Office of the President, Washington, D.C.

Walker, G. 1999. "Snowball Earth," *New Scientist,* 164(2211): 28–33.

Waller, G. (ed.) 1996. *Sea Life: A Complete Guide to the Marine Environment.* Smithsonian Institution Press, Washington, D.C.

Webster, P. J., and J. A. Curry. 1998. "The Oceans and Weather," *Scientific American Presents,* 9(3): 38–43.

Whynott, D. 1999. "The Most Expensive Fish in the Sea," *Discover,* 20(4): 80–85.

Wong, K. 1999. "Cetacean Creation," *Scientific American,* 280(1): 26–30.

Zieman, J. C. 1982. *The Ecology of the Seagrasses of South Florida: A Community Profile.* U.S. Fish and Wildlife Services/Office of Biological Services 82-25, Washington, D.C.

Index

Ellen J. Prager, PhD, is an experienced marine scientist who has taught oceanography for Sea Education Association, conducted research for the U.S. Geological Survey, is a Fellow of the Explorers Club, and has worked with organizations such as the National Geographic Society and MSNBC. She is a frequent speaker at scientific and educational conferences, has written several books and contributed to journals and magazines. During two week-long stays in the underwater laboratory, Aquarius 2000 in the Florida Keys, Prager recently studied coral reefs, was interviewed on *The Today Show* and broadcast live to students around the Nation. She is now a freelance writer and consultant living in Arlington, Virginia.

Sylvia A. Earle, PhD, called "Her Deepness" by the *New York Times* and *The New Yorker,* is the chair of Deep Ocean Exploration and Research and has been a spokesperson for SeaWeb. She was the first woman to serve as the chief scientist of the National Oceanic and Atmospheric Administration and is presently the "explorer-in-residence" at the National Geographic Society and leader of the Sustainable Seas Expedition.